T0305654

Chemotaxis Modeling of Autoimmune Inflammation

In response to an infection (e.g., from pathogens such as bacteria and viruses), the immune system can deplete macrophages (specialized white blood cells) and produce cytokines that are pro-inflammatory or anti-inflammatory. This counterproductive autoimmune response is represented mathematically as nonlinear chemotaxis diffusion.

This book is directed to the computer-based modeling of chemotaxis inflammation. The spatiotemporal analysis is based on a model of three partial differential equations (PDEs).

The three PDE model is coded (programmed) as a set of routines in R, a quality, open-source, scientific programming system. The numerical integration (solution) of the PDEs is by the method of lines (MOL).

The three PDE model can be used for computer-based experimentation, e.g., parameter variation and changes in the model equations or alternate models, to enhance a quantitative understanding of a postulated inflammation.

This experimentation is illustrated by chapters pertaining to: (1) the computation and display of the PDE time derivatives, (2) the RHS terms of the PDEs with emphasis on the chemotaxis terms, (3) parameter variations to demonstrate parameter effects and sensitivities, and (4) additional terms in the PDEs to include PDE coupling and extensions of the basic PDE model.

Chemotaxis Modeling of Autoimmune Inflammation

PDE Computer Analysis in R

William E. Schiesser

CRC Press
Taylor & Francis Group
Boca Raton London New York

CRC Press is an imprint of the
Taylor & Francis Group, an **informa** business

First edition published 2023
by CRC Press
6000 Broken Sound Parkway NW, Suite 300, Boca Raton, FL 33487-2742

and by CRC Press
4 Park Square, Milton Park, Abingdon, Oxon, OX14 4RN

CRC Press is an imprint of Taylor & Francis Group, LLC

ISBN: 978-1-032-31606-2 (hbk)
ISBN: 978-1-032-31764-9 (pbk)
ISBN: 978-1-003-31120-1 (ebk)

DOI: 10.1201/9781003311201

Typeset in Nimbus font
by KnowledgeWorks Global Ltd.

Contents

Preface

In response to an infection (e.g., from pathogens such as bacteria and viruses), the immune system can deplete macrophages (specialized white blood cells) and produce cytokines that are pro-inflammatory or anti-inflammatory. This counterproductive response is represented mathematically as nonlinear chemotaxis diffusion.

This book is directed to the computer-based modeling of *chemotaxis inflammation* [1]. The spatiotemporal analysis is based on a model of three partial differential equations (PDEs). The three PDE dependent and two independent variables are explained as follows:

Variables	Interpretation
m	concentration of macrophages
c	concentration of pro-inflammatory cytokines
a	concentration of anti-inflammatory cytokines
r	radial position in spherical coordinates
t	time

The three PDE model is coded (programmed) as a set of routines in R, a quality, open-source, scientific programming system [2]. r, t are the PDE independent variables. r is the radial coordinate in spherical coordinates (r, φ, θ) reflecting the geometry of infected tissue.

The three PDE model can be used for computer-based experimentation, e.g., parameter variation and changes in the model equations or alternate models, to enhance a quantitative understanding of a postulated inflammation.

This experimentation is illustrated by chapters pertaining to: (1) the computation and display of the PDE time derivatives, (2) the RHS terms of the PDEs with emphasis on the chemotaxis terms, (3) parameter variations to demonstrate parameter effects and sensitivities, and (4) additional terms in the PDEs to include PDE coupling and extensions of the basic PDE model.

The author would welcome reader impressions of the computer-based autoimmune inflammation model discussed in this book. Comments can be directed to wes1@lehigh.edu.

References

1. Giunta, M., Carmela Lombardo, and M. Sammartino (2021), Pattern Formation and Transition to Chaos in a Chemotaxis Model of Acute Inflammation. *SIAM Journal on Applied Dynamical Systems*, **20**, no. 4, pp. 1844–1881.

2. Soetaert, K., J. Cash, and F. Mazzia (2012), *Solving Differential Equations in R*, Springer-Verlag, Heidelberg, Germany.

1 PDE Chemotaxis Model Formulation

1.1 Introduction

In response to an infection (e.g., from pathogens such as bacteria and viruses), the immune system can deplete macrophages (specialized white blood cells) and produce cytokines that are pro-inflammatory or anti-inflammatory. This counterproductive (autoimmune) response is modeled in space and time as a system of partial differential equations (PDEs).

The PDE model is integrated (solved) numerically in R routines[1] by the method of lines (MOL)[2]. Basic R utilities present the spatiotemporal solution from the PDE model in numerical and graphical form.

1.2 Three PDE model coordinate-free formulation

In the following analysis, the macrophage cell, pro-inflammatory and anti-inflammatory cytokine concentrations are modeled as a system of three PDEs, starting with a coordinate-free PDE model formulation [1].

$$\frac{\partial m}{\partial t} = D\nabla^2 m - \nabla \cdot \left(\chi \frac{m}{(1+\alpha c)^2} \nabla c \right) + r_m mc(1-m) \tag{1.1-1}$$

$$\frac{\partial c}{\partial t} = \nabla^2 c + \frac{m}{1+\beta a^p} - c \tag{1.1-2}$$

$$\frac{\partial a}{\partial t} = \frac{1}{\tau} \left(\nabla^2 a + \frac{m}{1+\beta a^p} - a \right) \tag{1.1-3}$$

The dependent and independent variables, coordinate-free spatial operators, volumetric rate functions, and parameters of eqs. (1.1-1–1.1-3) are explained in Tables 1.1a, 1.1b.

Eqs. (1.1-1–1.1-3) are second order in space, and therefore each requires two boundary conditions (BCs) that are taken as homogeneous (zero derivative, zero flux) Neumann BCs [1].

$$\nabla m = \nabla c = \nabla a = 0 \text{ on } \partial \Omega \tag{1.2-1}$$

where Ω is the spatial domain and $\partial \Omega$ is the boundary where the BCs apply.

---

[1] The coding (programming) is in R, a quality, open-source, scientific programming system that is readily available on the Internet [2].

[2] The method of lines is a general numerical algorithm for PDEs in which the boundary value (spatial) derivatives are replaced with algebraic approximations; in this case, finite differences (FDs). The resulting system of initial value ordinary differential equations (ODEs) is then integrated (solved) with a library ODE integrator.

DOI: 10.1201/9781003311201-1

1

Table 1.1a
Variables in eqs. (1.1-1–1.1-3)

Variables	Interpretation
m	concentration of macrophages
c	concentration of pro-inflammatory cytokines
a	concentration of anti-inflammatory cytokines
t	time

Table 1.1b
Operators in eqs. (1.1-1–1.1-3)

Operators	Interpretation (Based on Differential Volume for Selected Spatial Coordinate System)
$\nabla\cdot$	divergence of a vector
∇	gradient of a scalar

Table 1.1c
Rate Functions in eqs. (1.1-1–1.1-3)

Rate Functions	Interpretation (Based on Differential Volume for Selected Spatial Coordinate System)
$\dfrac{\partial m}{\partial t}, \dfrac{\partial c}{\partial t}, \dfrac{\partial a}{\partial t}$	addition (> 0) or depletion (< 0) of m, c, a
$D\nabla^2 m, \nabla^2 c, \nabla^2 a/\tau$	linear (Fick's first law) diffusion of m, c, a
$-\nabla \cdot \left(\chi \dfrac{m}{(1+\alpha c)^2} \nabla c \right)$	chemotaxis of m
$r_m mc(1-m)$	activation of m
$\dfrac{m}{1+\beta a^p}$	production of c
$-c$	decay of c
$\dfrac{1}{\tau}\left(\dfrac{m}{1+\beta a^p} \right)$	production of a
$-\dfrac{a}{\tau}$	decay of a

Table 1.1d

Parameters in eqs. (1.1-1–1.1-3)

Parameters	Interpretation
D	dimensionless diffusivity
χ, α	chemotaxis parameters
r_m	rate constant for activation of m
β, p	parameters in rate of production function of c, a
τ	time scale adjustment for a

Eqs. (1.1-1–1.1-3) are first order in time, and each requires one initial condition (IC). The three ICs are specified once the spatial domain is defined.

1.3 Three PDE model formulation in spherical coordinates

If the spatial domain is defined in spherical coordinates $(r\phi, \theta)$, the operators in Tables 1.2a, 1.2b are

Table 1.2a

∇· (Divergence of a Vector, Spherical Coordinates)

$$
\begin{bmatrix}
[\nabla]_r = \dfrac{1}{r^2} \dfrac{\partial}{\partial r}(r^2\ ) \\[2ex]
[\nabla]_\theta = \dfrac{1}{r\sin\theta} \dfrac{\partial}{\partial \theta}(\sin\theta\ ) \\[2ex]
[\nabla]_\phi = \dfrac{1}{r\sin\theta} \dfrac{\partial}{\partial \phi}
\end{bmatrix}
$$

Table 1.2b

∇ (Gradient of a Scalar, Spherical Coordinates)

$$
\begin{bmatrix}
[\nabla]_r = \dfrac{\partial}{\partial r} \\[2ex]
[\nabla]_\theta = \dfrac{1}{r} \dfrac{\partial}{\partial \theta} \\[2ex]
[\nabla]_\phi = \dfrac{1}{r\sin\theta} \dfrac{\partial}{\partial \phi}
\end{bmatrix}
$$

If angular variations are neglected, $(r, \phi, \theta) \to r$, the linear diffusion operator in eqs. (1.1-1–1.1-3) is

$$\nabla \cdot = \frac{1}{r^2} \frac{\partial}{\partial r}(r^2)$$

$$\nabla = \frac{\partial}{\partial r}$$

$$\nabla^2 = \nabla \cdot \nabla = \frac{1}{r^2} \frac{\partial \left(r^2 \frac{\partial}{\partial r} \right)}{\partial r} = \frac{\partial^2}{\partial r^2} + \frac{2}{r} \frac{\partial}{\partial r}$$

The second term is indeterminate at $r = 0$ and can be resolved by l'Hospital's rule.

$$\frac{2}{r} \frac{\partial}{\partial r} \Big|_{r \to 0} = 2 \frac{\partial^2}{\partial r^2}$$

The chemotaxis term in spherical coordinates is

$$\nabla \cdot \left(\chi \frac{m}{(1+\alpha c)^2} \nabla c \right) = \frac{1}{r^2} \frac{\partial}{\partial r}(r^2) \left(\chi \frac{m}{(1+\alpha c)^2} \frac{\partial c}{\partial r} \right)$$

$$= \frac{\partial \left(\chi \frac{m}{(1+\alpha c)^2} \right)}{\partial r} \frac{\partial c}{\partial r} + \chi \frac{m}{(1+\alpha c)^2} \frac{\partial^2 c}{\partial r^2} + \chi \frac{m}{(1+\alpha c)^2} \frac{2}{r} \frac{\partial c}{\partial r}$$

At $r = 0$, this term is

$$\nabla \cdot \left(\chi \frac{m}{(1+\alpha c)^2} \nabla c \right) \Big|_{r \to 0}$$

$$= \frac{\partial \left(\chi \frac{m}{(1+\alpha c)^2} \right)}{\partial r} \frac{\partial c}{\partial r} \Big|_{r \to 0} + \chi \frac{m}{(1+\alpha c)^2} \frac{\partial^2 c}{\partial r^2} \Big|_{r \to 0} + \chi \frac{m}{(1+\alpha c)^2} 2 \frac{\partial^2 c}{\partial r^2}$$

$$= 3\chi \frac{m}{(1+\alpha c)^2} \frac{\partial^2 c}{\partial r^2}$$

The PDEs for $\dfrac{\partial m(r > 0, t)}{\partial t}$, $\dfrac{\partial c(r > 0, t)}{\partial t}$, $\dfrac{\partial a(r > 0, t)}{\partial t}$ can now be stated.

$$\frac{\partial m}{\partial t} = D \left(\frac{\partial^2 m}{\partial r^2} + \frac{2}{r} \frac{\partial m}{\partial r} \right)$$

$$- \frac{\partial \chi \frac{m}{(1+\alpha c)^2}}{\partial r} \frac{\partial c}{\partial r} - \chi \frac{m}{(1+\alpha c)^2} \frac{\partial^2 c}{\partial r^2} - \chi \frac{m}{(1+\alpha c)^2} \frac{2}{r} \frac{\partial c}{\partial r}$$

$$+ r_m m c (1 - m) \tag{1.3-1}$$

$$\frac{\partial c}{\partial t} = \frac{\partial^2 c}{\partial r^2} + \frac{2}{r}\frac{\partial c}{\partial r} + \frac{m}{1+\beta a^p} - c \tag{1.3-2}$$

$$\frac{\partial a}{\partial t} = \frac{1}{\tau}\left(\frac{\partial^2 a}{\partial r^2} + \frac{2}{r}\frac{\partial a}{\partial r} + \frac{m}{1+\beta a^p} - a\right) \tag{1.3-3}$$

Eqs. (1.3-1–1.3-6) for $\dfrac{\partial m(r=0,t)}{\partial t}$, $\dfrac{\partial c(r=0,t)}{\partial t}$, $\dfrac{\partial a(r=0,t)}{\partial t}$ are (with BCs (1.4-1–1.4-3))

$$\frac{\partial m}{\partial t} = 3D\frac{\partial^2 m}{\partial r^2} - 3\chi\frac{m}{(1+\alpha c)^2}\frac{\partial^2 c}{\partial r^2} + r_m mc(1-m) \tag{1.3-4}$$

$$\frac{\partial c}{\partial t} = 3\frac{\partial^2 c}{\partial r^2} + \frac{m}{1+\beta a^p} - c \tag{1.3-5}$$

$$\frac{\partial a}{\partial t} = \frac{1}{\tau}\left(3\frac{\partial^2 a}{\partial r^2} + \frac{m}{1+\beta a^p} - a\right) \tag{1.3-6}$$

The BCs for eqs. (1.3-1–1.3-6) are

$$\frac{\partial m(r=r_l=0,t)}{\partial r} = \frac{\partial m(r=r_u=1,t)}{\partial r} = 0 \tag{1.4-1}$$

$$\frac{\partial c(r=r_l=0,t)}{\partial r} = \frac{\partial c(r=r_u=1,t)}{\partial r} = 0 \tag{1.4-2}$$

$$\frac{\partial a(r=r_l=0,t)}{\partial r} = \frac{\partial a(r=r_u=1,t)}{\partial r} = 0 \tag{1.4-3}$$

Eqs. (1.3-1–1.3-6) and (1.4-1–1.4-3) constitute the chemotaxis model that is coded (implemented, programmed) in Chapter 2 as a set of R routines [2].

Summary and conclusion

The chemotaxis PDE model that leads to inflammation from reduction of the macrophage population (concentration) is stated as eqs. (1.3-1–1.3-6) and (1.4-1–1.4-3). The ICs are specified in the R main program discussed in Chapter 2.

References

1. Giunta, M., Carmela Lombardo, and M. Sammartino (2021), Pattern Formation and Transition to Chaos in a Chemotaxis Model of Acute Inflammation. *SIAM Journal on Applied Dynamical Systems*, **20**, no. 4, pp. 1844–1881.

2. Soetaert, K., J. Cash, and F. Mazzia (2012), *Solving Differential Equations in R*, Springer-Verlag, Heidelberg, Germany.

2 PDE Chemotaxis Model Implementation

2.1 Introduction

The computer implementation of the chemotaxis model [1] defining the macrophage, pro-inflammatory cytokine, and anti-inflammatory cytokine concentrations, $m(r,t)$, $c(r,t)$, $a(r,t)$, of eqs. (1.3-1–1.3-6) is discussed in this chapter. The coding (programming) is in R, a quality, open-source, scientific programming system that is readily available on the Internet [2]. The main program for the model is listed next.

2.2 Coding of the chemotaxis model

The following main program for eqs. (1.3-1–1.3-6) is based on the use of the basic R system [2].

2.2.1 Main program

```
#
# Three PDE chemotaxis inflammation model
#
# Delete previous workspaces
  rm(list=ls(all=TRUE))
#
# Access ODE integrator
  library("deSolve");
#
# Access functions for numerical solution
  setwd("f:/inflammation/chap2");
  source("pde1a.R");
  source("dss004.R");
#
# Select case
  ncase=1;
#
# Parameters
  D=1;
  rm=1;
  chi=1;
  alpha=1;
  beta=1;
```

DOI: 10.1201/9781003311201-2

```
  p=1;
  tau=1;
  m0=1;
#
# Grid (in r)
  nr=21;rl=0;ru=1
  r=seq(from=rl,to=ru,by=(ru-rl)/(nr-1));
#
# Independent variable for ODE integration
  t0=0;tf=1;nout=11;
  tout=seq(from=t0,to=tf,by=(tf-t0)/(nout-1));
#
# Initial condition
  u0=rep(0,3*nr);
  if(ncase==1){
    for(ir in 1:nr){
      u0[ir]      =m0;
      u0[ir+nr] =0;
      u0[ir+2*nr]=0;
    }
  }
  if(ncase==2){
    for(ir in 1:nr){
      u0[ir]      =0.5*(1+cos(pi*(r[ir]-rl)/(ru-rl)));
      u0[ir+nr] =0;
      u0[ir+2*nr]=0;
    }
  }
  ncall=0;
#
# ODE integration
  out=lsodes(y=u0,times=tout,func=pde1a,
       sparsetype ="sparseint",rtol=1e-6,
       atol=1e-6,maxord=5);
  nrow(out)
  ncol(out)
#
# Arrays for plotting numerical solution
  m=matrix(0,nrow=nr,ncol=nout);
  c=matrix(0,nrow=nr,ncol=nout);
  a=matrix(0,nrow=nr,ncol=nout);
  for(it in 1:nout){
    for(ir in 1:nr){
      m[ir,it]=out[it,ir+1];
      c[ir,it]=out[it,ir+1+nr];
```

```
        a[ir,it]=out[it,ir+1+2*nr];
    }
  }
#
# Display numerical solution
  for(it in 1:nout){
    if((it==1)|(it==nout)){
      cat(sprintf("\n    t      r       m(r,t)       c(r,t)
        a(r,t)\n"));
      for(ir in 1:nr){
        cat(sprintf("%6.2f%6.2f%12.3e%12.3e%12.3e\n",
          tout[it],r[ir],m[ir,it],c[ir,it],a[ir,it]));
      }
    }
  }
#
# Calls to ODE routine
  cat(sprintf("\n\n ncall = %5d\n\n",ncall));
#
# Plot PDE solutions
#
# m
  par(mfrow=c(1,1));
  if(ncase==1){
  matplot(x=r,y=m,type="l",xlab="r",ylab="m(r,t)",
        xlim=c(rl,ru),lty=1,main="m(r,t)",lwd=2,
        col="black",ylim=c(0.9*m0,1.1*m0));}
  if(ncase==2){
  matplot(x=r,y=m,type="l",xlab="r",ylab="m(r,t)",
        xlim=c(rl,ru),lty=1,main="m(r,t)",lwd=2,
        col="black");
  persp(r,tout,m,theta=60,phi=45,
       xlim=c(rl,ru),ylim=c(t0,tf),zlim=c(0,1.1),
       xlab="r",ylab="t",zlab="m(r,t)");
  }
#
# c
  par(mfrow=c(1,1));
  matplot(x=r,y=c,type="l",xlab="r",ylab="c(r,t)",
        xlim=c(rl,ru),lty=1,main="c(r,t)",lwd=2,
        col="black");
#
# a
  par(mfrow=c(1,1));
  matplot(x=r,y=a,type="l",xlab="r",ylab="a(r,t)",
```

```
      xlim=c(rl,ru),lty=1,main="a(r,t)",lwd=2,
      col="black");
```

Listing 2.1   Main program for eqs. (1.3-1–1.3-6)

We can note the following details about Listing 2.1.
- Previous workspaces are deleted.

```
#
# Three PDE chemotaxis inflammation model
#
# Delete previous workspaces
  rm(list=ls(all=TRUE))
```

- The R ODE integrator library deSolve is accessed [2]. Then the directory with the files for the solution of eqs. (1.3-1–1.3-6) is designated. Note that setwd (set working directory) uses / rather than the usual \.

```
#
# Access ODE integrator
  library("deSolve");
#
# Access functions for numerical solution
  setwd("f:/inflammation/chap2");
  source("pde1a.R");
  source("dss004.R");
  source("dss044.R");
```

dss004, dss044 are library routines for the calculation of spatial first and second derivatives. These routines are listed and explained in Appendix A1.
- A case is selected for the initial condition (IC).

```
#
# Select case
  ncase=1;
```

- The model parameters are specified numerically.

```
#
# Parameters
  D=1;
  rm=1;
  chi=1;
  alpha=1;
  beta=1;
  p=1;
  tau=1;
  m0=1;
```

The parameters are listed in Table 1.1d.

- A spatial grid for eqs. (1.3-1–1.3-6) is defined with 21 points so that r = 0,0.05,...,1.

```
#
# Grid (in r)
  nr=21;rl=0;ru=1
  r=seq(from=rl,to=ru,by=(ru-rl)/(nr-1));
```

- An interval in *t* is defined for 11 output points, so that tout=0,0.1,...,1.

```
#
# Independent variable for ODE integration
  t0=0;tf=1;nout=11;
  tout=seq(from=t0,to=tf,by=(tf-t0)/(nout-1));
```

- For ncase=1, the ICs for eqs. (1.3-1–1.3-6) are constant in *r*. Also, the counter for the calls to pde1a is initialized.
- For ncase=2, the IC for eqs. (1.3-1, 1.3-4) is a half cos wave and the ICs for eqs. (1.3-2, 1.3-3, 1.3-5, 1.3-6) are zero.
  if(ncase==2) for(ir in 1:nr) u0[ir] =0.5*(1+cos(pi*(r[ir]-rl)/(ru-rl)));
  u0[ir+nr] =0; u0[ir+2*nr]=0;
- The system of 3(21) = 63 ODEs is integrated by the library integrator lsodes (available in deSolve, [2]). As expected, the inputs to lsodes are the ODE/MOL function, pde1a, the IC vector, u0, and the vector of output values of *t*, tout. The length of u0 (63) informs lsodes how many ODEs are to be integrated. func,y,times are reserved names.

```
#
# ODE integration
  out=lsodes(y=u0,times=tout,func=pde1a,
      sparsetype ="sparseint",rtol=1e-6,
      atol=1e-6,maxord=5);
  nrow(out)
  ncol(out)
```

nrow,ncol confirm the dimensions of out.
- $m(r,t), c(r,t), a(r,t)$ are placed in matrices for subsequent plotting.

```
#
# Arrays for plotting numerical solution
   m=matrix(0,nrow=nr,ncol=nout);
   c=matrix(0,nrow=nr,ncol=nout);
   a=matrix(0,nrow=nr,ncol=nout);
  for(it in 1:nout){
    for(ir in 1:nr){
       m[ir,it]=out[it,ir+1];
       c[ir,it]=out[it,ir+1+nr];
```

```
        a[ir,it]=out[it,ir+1+2*nr];
      }
    }
```

The offset +1 is required because the first element of the solution vectors in out is the value of t and the 2 to 64 elements are the 63 values of $m(r,t), c(r,t), a(r,t)$. These dimensions from the preceding calls to
nrow,ncol are confirmed in the subsequent output.

- The numerical values of $m(r,t), c(r,t), a(r,t)$ returned by lsodes are
  displayed.

```
#
# Display numerical solution
  for(it in 1:nout){
    if((it==1)|(it==nout)){
      cat(sprintf("\n    t       r        m(r,t)         c(r,t)
        a(r,t)\n"));
      for(ir in 1:nr){
        cat(sprintf("%6.2f%6.2f%12.3e%12.3e%12.3e\n",
            tout[it],r[ir],m[ir,it],c[ir,it],a[ir,it]));
      }
    }
  }
```

- The number of calls to pde1a is displayed at the end of the solution.

```
#
# Calls to ODE routine
  cat(sprintf("\n\n ncall = %5d\n\n",ncall));
```

- $m(r,t), c(r,t), a(r,t)$ are plotted against r and parametrically in t with the
  R utility matplot. par(mfrow=c(1,1)) specifies a 1×1 matrix of plots,
  that is, one plot on a page.

```
#
# Plot PDE solutions
#
# m
  par(mfrow=c(1,1));
  if(ncase==1){
  matplot(x=r,y=m,type="l",xlab="r",ylab="m(r,t)",
          xlim=c(rl,ru),lty=1,main="m(r,t)",lwd=2,
          col="black",ylim=c(0.9*m0,1.1*m0));}
  if(ncase==2){
  matplot(x=r,y=m,type="l",xlab="r",ylab="m(r,t)",
          xlim=c(rl,ru),lty=1,main="m(r,t)",lwd=2,
```

```
                col="black");
      persp(r,tout,m,theta=60,phi=45,
              xlim=c(rl,ru),ylim=c(t0,tf),zlim=c(0,1.1),
              xlab="r",ylab="t",zlab="m(r,t)");
      }
  #
  # c
    par(mfrow=c(1,1));
    matplot(x=r,y=c,type="l",xlab="r",ylab="c(r,t)",
              xlim=c(rl,ru),lty=1,main="c(r,t)",lwd=2,
              col="black");
  #
  # a
    par(mfrow=c(1,1));
    matplot(x=r,y=a,type="l",xlab="r",ylab="a(r,t)",
              xlim=c(rl,ru),lty=1,main="a(r,t)",lwd=2,
              col="black");
```

For ncase=1 scaling of the y-axis, ylim=c(0.9*m0,1.1*m0), is used since the departure of $m(r,t)$ from the IC m_0 is so small, plotting by matplot with y-axis automatic scaling fails with an error message.
For ncase=2 a call to the R utility persp is included to provide a 3D plot of $m(r,t)$.

This completes the discussion of the main program for eqs. (1.3-1–1.3-6). The ODE/MOL routine pde1a called by lsodes from the main program for the numerical MOL integration of eqs. (1.3-1–1.3-6) is discussed next.

2.2.2   ODE/MOL routine

pde1a called in the main program of Listing 2.1 follows.

```
  pde1a=function(t,u,parms){
#
# Function pde1a computes the t derivative
# vectors of m(r,t),c(r,t),a(r,t)
#
# One vector to three vectors
  m=rep(0,nr);c=rep(0,nr);a=rep(0,nr);
  for(ir in 1:nr){
    m[ir]=u[ir];
    c[ir]=u[ir+nr];
    a[ir]=u[ir+2*nr];
  }
#
# mr,cr,ar
```

```
   mr=dss004(rl,ru,nr,m);
   cr=dss004(rl,ru,nr,c);
   ar=dss004(rl,ru,nr,a);
#
# BCs
   mr[1]=0;mr[nr]=0;
   cr[1]=0;cr[nr]=0;
   ar[1]=0;ar[nr]=0;
#
# mrr,crr,arr
   nl=2;nu=2;
   mrr=dss044(rl,ru,nr,m,mr,nl,nu);
   crr=dss044(rl,ru,nr,c,cr,nl,nu);
   arr=dss044(rl,ru,nr,a,ar,nl,nu);
#
# Functions of m,c,a
   func1=rep(0,nr);
   func2=rep(0,nr);
   for(ir in 1:nr){
     func1[ir]=chi*m[ir]/(1+alpha*c[ir])^2;
     func2[ir]=m[ir]/(1+beta*a[ir]^p);
   }
   dfunc1=dss004(rl,ru,nr,func1);
#
# PDEs
   mt=rep(0,nr);ct=rep(0,nr);at=rep(0,nr);
   for(ir in 1:nr){
     if(ir==1){
       mt[ir]=3*D*mrr[ir]-3*func1[ir]*crr[ir]+rm*m[ir]
         *c[ir]*(1-m[ir]));
       ct[ir]=3*crr[ir]+func2[ir]-c[ir];
       at[ir]=(1/tau)*(3*arr[ir]+func2[ir]-a[ir]);
     }
    if(ir==nr){
       mt[ir]=D*mrr[ir]-func1[ir]*crr[ir]+rm*m[ir]*c[ir]
         *(1-m[ir]));
       ct[ir]=crr[ir]+func2[ir]-c[ir];
       at[ir]=(1/tau)*(arr[ir]+func2[ir]-a[ir]);
     }
     if((ir>1)&(ir<nr)){
       mt[ir]=D*(mrr[ir]+(2/r[ir])*mr[ir])-
         dfunc1[ir]*cr[ir]-func1[ir]*crr[ir]-func1[ir]
           *(2/r[ir])*
         cr[ir]+rm*m[ir]*c[ir]*(1-m[ir]));
       ct[ir]=(crr[ir]+(2/r[ir])*cr[ir])+func2[ir]-c[ir];
```

```
   at[ir]=(1/tau)*(arr[ir]+(2/r[ir])*ar[ir]+func2[ir]
     -a[ir]);
  }
 }
#
# Three vectors to one vector
 ut=rep(0,3*nr);
 for(ir in 1:nr){
   ut[ir]     =mt[ir];
   ut[ir+nr] =ct[ir];
   ut[ir+2*nr]=at[ir];
 }
#
# Increment calls to pde1a
 ncall <<- ncall+1;
#
# Return derivative vector
 return(list(c(ut)));
 }
```

Listing 2.2   ODE/MOL routine for eqs. (1.3-1–1.3-6)

We can note the following details about Listing 2.2.

- The function is defined.

```
  pde1a=function(t,u,parms){
#
# Function pde1a computes the t derivative
# vectors of m(r,t),c(r,t),a(r,t)
```

  t is the current value of t in eqs. (1.3-1–1.3-6). u is the 63-vector of ODE/PDE dependent variables. parm is an argument to pass parameters to pde1a (unused, but required in the argument list). The arguments must be listed in the order stated to properly interface with lsodes called in the main program of Listing 2.1. The derivative vector of the LHS of eqs. (1.3-1–1.3-6) is calculated and returned to lsodes as explained subsequently.
- Vector u is placed in three vectors to facilitate the programming of eqs. (1.3-1–1.3-6).

```
#
# One vector to three vectors
  m=rep(0,nr);c=rep(0,nr);a=rep(0,nr);
  for(ir in 1:nr){
    m[ir]=u[ir];
    c[ir]=u[ir+nr];
```

```
    a[ir]=u[ir+2*nr];
  }
```

- The first spatial derivatives $\frac{\partial m(r,t)}{\partial r}\ \frac{\partial c(r,t)}{\partial r}\ \frac{\partial a(r,t)}{\partial r}$ are computed by dss004, a library routine for first order spatial derivatives. dss004 is listed in Appendix A1 with an explanation of the arguments.

```
#
# mr,cr,ar
    mr=dss004(rl,ru,nr,m);
    cr=dss004(rl,ru,nr,c);
    ar=dss004(rl,ru,nr,a);
```

- Eqs. (1.4-1–1.4-3) are programmed (homogeneous, no flux Neumann BCs).

```
#
# BCs
    mr[1]=0;mr[nr]=0;
    cr[1]=0;cr[nr]=0;
    ar[1]=0;ar[nr]=0;
```

Subscripts 1,nr correspond to $r=r_l=0, r=r_u=1$.

- The second spatial derivatives $\frac{\partial^2 m(r,t)}{\partial r^2}\ \frac{\partial^2 c(r,t)}{\partial r^2}\ \frac{\partial^2 a(r,t)}{\partial r^2}$ are computed by dss044, a library routine for second order spatial derivatives. dss044 is listed in Appendix A1 with an explanation of the arguments.

```
#
# mrr,crr,arr
    nl=2;nu=2;
    mrr=dss044(rl,ru,nr,m,mr,nl,nu);
    crr=dss044(rl,ru,nr,c,cr,nl,nu);
    arr=dss044(rl,ru,nr,a,ar,nl,nu);
```

nl=2,nu=2 specify Neumann BCs at $r=rl,r_u$.

- The functions $\frac{m}{(1+\alpha c)^2}, \frac{m}{1+\beta a^p}\ \frac{\partial \chi \frac{m}{(1+\alpha c)^2}}{\partial r}$ in eqs. (1.3-1–1.3-6) are computed as func1,func2,dfunc1.

```
#
# Functions of m,c,a
    func1=rep(0,nr);
    func2=rep(0,nr);
    for(ir in 1:nr){
      func1[ir]=chi*m[ir]/(1+alpha*c[ir])^2;
      func2[ir]=m[ir]/(1+beta*a[ir]^p);
    }
    dfunc1=dss004(rl,ru,nr,func1);
```

- Eqs. (1.3-1–1.3-6) are programmed in the MOL format as a system of 3*nr=63 approximating ODEs.

```
#
# PDEs
  mt=rep(0,nr);ct=rep(0,nr);at=rep(0,nr);
  for(ir in 1:nr){
    if(ir==1){
      mt[ir]=3*D*mrr[ir]-3*func1[ir]*crr[ir]+rm*m[ir]*c[ir]
        *(1-m[ir]);
      ct[ir]=3*crr[ir]+func2[ir]-c[ir];
      at[ir]=(1/tau)*(3*arr[ir]+func2[ir]-a[ir]);
    }
    if(ir==nr){
      mt[ir]=D*mrr[ir]-func1[ir]*crr[ir]+rm*m[ir]*c[ir]
        *(1-m[ir]);
      ct[ir]=crr[ir]+func2[ir]-c[ir];
      at[ir]=(1/tau)*(arr[ir]+func2[ir]-a[ir]);
    }
    if((ir>1)&(ir<nr)){
      mt[ir]=D*(mrr[ir]+(2/r[ir])*mr[ir])-
        dfunc1[ir]*cr[ir]-func1[ir]*crr[ir]-func1[ir]
        *(2/r[ir])*cr[ir]+
        rm*m[ir]*c[ir]*(1-m[ir]);
      ct[ir]=(crr[ir]+(2/r[ir])*cr[ir])+func2[ir]-c[ir];
      at[ir]=(1/tau)*(arr[ir]+(2/r[ir])*ar[ir]+func2[ir]
        -a[ir]);
    }
  }
```

This code requires some additional explanation.

- Arrays are defined for the t derivatives, $\dfrac{\partial m}{\partial t}$, $\dfrac{\partial c}{\partial t}$, $\dfrac{\partial a}{\partial t}$, of eqs. (1.3-1–1.3-6).

```
#
# PDEs
  mt=rep(0,nr);ct=rep(0,nr);at=rep(0,nr);
```

- The MOL ODEs are programmed over the interval $r_l \leq r \leq r_u$ with a for.

```
  for(ir in 1:nr){
    if(ir==1){
      mt[ir]=3*D*mrr[ir]-3*func1[ir]*crr[ir]+rm*m[ir]
        *c[ir]*(1-m[ir]);
```

For $r = r_l$, eq. (1.3-4) is programmed using previously computed terms.

- Eq. (1.3-5) is programmed as

```
ct[ir]=3*crr[ir]+func2[ir]-c[ir];
```

- Eq. (1.3-6) is programmed as

```
at[ir]=(1/tau)*(3*arr[ir]+func2[ir]-a[ir]);
}
```

- For $r = r_u$, eq. (1.3-1) is programmed as

```
if(ir==nr){
  mt[ir]=D*mrr[ir]-func1[ir]*crr[ir]+rm*m[ir]*c[ir]
  *(1-m[ir]);
```

Index nr corresponds to $r = r_u$. BC (1.4-2) is included.
- For $r = r_u$, eq. (1.3-2) is programmed as

```
ct[ir]=crr[ir]+func2[ir]-c[ir];
```

BC (1.4-2) is included.
- For $r = r_u$, eq. (1.3-3) is programmed as

```
at[ir]=(1/tau)*(arr[ir]+func2[ir]-a[ir]);
```

BC (1.4-3) is included.
- For $r_l \leq r \leq r_u$, eq. (1.3-1) is programmed as

```
if((ir>1)&(ir<nr)){
  mt[ir]=D*(mrr[ir]+(2/r[ir])*mr[ir])-
    dfunc1[ir]*cr[ir]-func1[ir]*crr[ir]-func1[ir]
    *(2/r[ir])*
    cr[ir]+rm*m[ir]*c[ir]*(1-m[ir]);
```

- For $r_l \leq r \leq r_u$, eq. (1.3-2) is programmed as

```
ct[ir]=(crr[ir]+(2/r[ir])*cr[ir])+func2[ir]-c[ir];
```

- For $r_l \leq r \leq r_u$, eq. (1.3-3) is programmed as

```
at[ir]=(1/tau)*(arr[ir]+(2/r[ir])*ar[ir]
  +func2[ir]-a[ir]);
}
}
```

The concluding }s end $r_l \leq r \leq r_u$ and the for in r.

- The 63 ODE derivatives are placed in the vector ut for return to lsodes to take the next in t along the solution.

```
#
# Three vectors to one vector
ut=rep(0,3*nr);
for(ir in 1:nr){
```

```
        ut[ir]      =mt[ir];
        ut[ir+nr]   =ct[ir];
        ut[ir+2*nr]=at[ir];
      }
```

- The counter for the calls to pde1a is incremented and returned to the main program of Listing 2.1 by <<-.

```
#
# Increment calls to pde1a
  ncall <<- ncall+1;
```

- The vector ut is returned as a list as required by lsodes. c is the R vector utility. The final } concludes pde1a.

```
#
# Return derivative vector
  return(list(c(ut)));
  }
```

The concluding } ends pde1a.

This completes the discussion of pde1a. The output from the main program of Listing 2.1 and ODE/MOL routine pde1a of Listing 2.2 is considered next.

2.2.3 Numerical, graphical output

Abbreviated output from the main program and ODE/MOL routines of Listings 2.1, 2.2 for ncase=1 is as follows (see Table 2.1).

We can note the following details about this output.

- The ICs for eqs. (1.3-1–1.3-6) are verified ($t = 0$). This verification is important since if the ICs are incorrect, the subsequent solution will be incorrect.
- The output is for $t = 0, 1$ as programmed in Listing 2.1.
- The output is for $r = 0, 1/(21 - 1) = 0.05, \ldots, 1$ as programmed in Listing 2.1 ($3(nr) + 1 = 3(21) + 1 = 64$) values at each t. The +1 reflects the value of t at each of the 63 ODE solutions in matrix out from lsodes.
- The values of $m(r,t), c(r,t), a(r,t)$ do not change with r reflecting the constant ICs and homogeneous Neumann BCs for ncase=1. This is an important check since variation of the solutions with r would indicate a programming error.
- $c(r,t)$ and $a(r,t)$ are the same since the MOL ODEs are the same for the ICs and parameter values in Listing 2.1, particularly tau=1 (see the programming for $\dfrac{\partial c}{\partial t}, \dfrac{\partial a}{\partial t}$ in pde1a of Listing 2.2).
- The computational effort is modest, ncall = 122, so lsodes calculates the solution efficiently.

Table 2.1
Numerical Output for eqs. (1.3-1–1.3-6) ncase=1

[1] 11

[1] 64

```
    t       r       m(r,t)       c(r,t)       a(r,t)
  0.00    0.00    1.000e+00    0.000e+00    0.000e+00
  0.00    0.05    1.000e+00    0.000e+00    0.000e+00
  0.00    0.10    1.000e+00    0.000e+00    0.000e+00
            .                     .
            .                     .
            .                     .
Output for t=0.15 to 0.85 is removed
            .                     .
            .                     .
            .                     .
  0.00    0.90    1.000e+00    0.000e+00    0.000e+00
  0.00    0.95    1.000e+00    0.000e+00    0.000e+00
  0.00    1.00    1.000e+00    0.000e+00    0.000e+00

    t       r       m(r,t)       c(r,t)       a(r,t)
  1.00    0.00    1.000e+00    4.774e-01    4.774e-01
  1.00    0.05    1.000e+00    4.774e-01    4.774e-01
  1.00    0.10    1.000e+00    4.774e-01    4.774e-01
            .                     .
            .                     .
            .                     .
Output for t=0.15 to 0.85 is removed
            .                     .
            .                     .
            .                     .
  1.00    0.90    1.000e+00    4.774e-01    4.774e-01
  1.00    0.95    1.000e+00    4.774e-01    4.774e-01
  1.00    1.00    1.000e+00    4.774e-01    4.774e-01

ncall =    122
```

---

m(r,t)

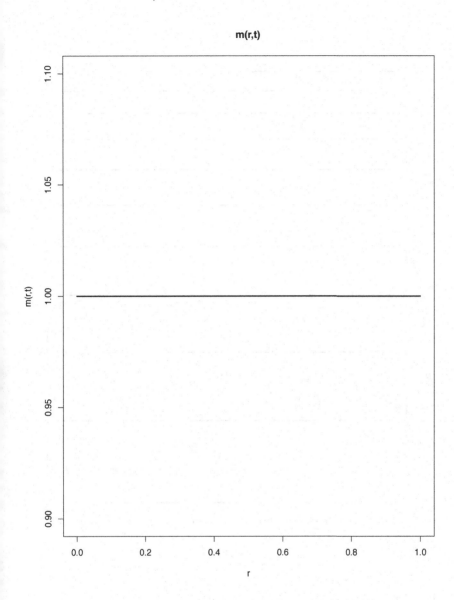

Figure 2.1-1 Numerical solution $m(r,t)$ from eqs. (1.3-1, 1.3-4), ncase=1.

The graphical output is in Figures 2.1-1–2.1-3.

Figure 2.1-1 indicates that $m(r,t)$ starts from the IC in Listing 2.1 and essentially remains at this value.

Figure 2.1-2 indicates that $c(r,t)$ is invariant in r and increases with t for ncase=1.

Figure 2.1-2 $c(r,t)$ from eqs. (1.3-2, 1.3-5), ncase=1.

Figure 2.1-3 indicates that $a(r,t)$ is invariant in r and increases with t for ncase=1.

$c(r,t), a(r,t)$ appear to be reaching an equilibrium (steady state) solution in Figs. 2.1-2, 2.1-3 (0.617). An equilibrium solution could be confirmed by increasing the final value of t, tf. This case is left as an exercise.

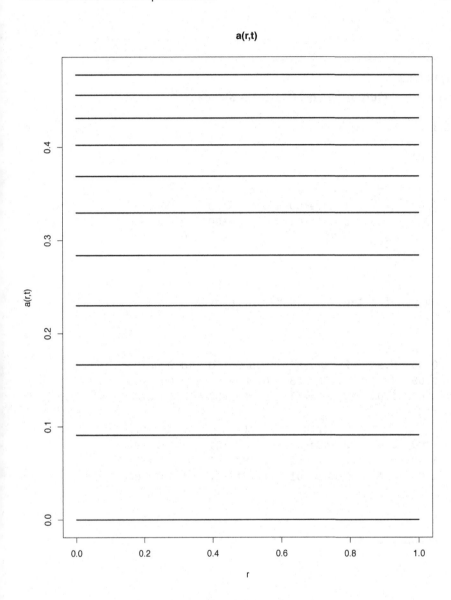

Figure 2.1-3 $a(r,t)$ from eqs. (1.3-3, 1.3-6), ncase=1.

Abbreviated output from the main program and ODE/MOL routines of Listings 2.1, 2.2 for ncase=2 is as follows (see Table 2.2).

We can note the following details about this output.

- The ICs for eqs. (1.3-1–1.3-6) are verified ($t = 0$). For example,

Table 2.2

Numerical Output for eqs. (1.3-1–1.3-6), ncase=2

[1] 11

[1] 64

t	r	m(r,t)	c(r,t)	a(r,t)
0.00	0.00	1.000e+00	0.000e+00	0.000e+00
0.00	0.05	9.938e-01	0.000e+00	0.000e+00
0.00	0.10	9.755e-01	0.000e+00	0.000e+00
	.	.	.	
	.	.	.	
	.	.	.	

Output for t=0.15 to 0.85 is removed

	.	.		
	.	.		
	.	.		
0.00	0.90	2.447e-02	0.000e+00	0.000e+00
0.00	0.95	6.156e-03	0.000e+00	0.000e+00
0.00	1.00	0.000e+00	0.000e+00	0.000e+00

t	r	m(r,t)	c(r,t)	a(r,t)
1.00	0.00	2.143e-01	1.182e-01	1.182e-01
1.00	0.05	2.142e-01	1.182e-01	1.182e-01
1.00	0.10	2.141e-01	1.181e-01	1.181e-01
	.	.		
	.	.		
	.	.		

Output for t=0.15 to 0.85 is removed

	.	.		
	.	.		
	.	.		
1.00	0.90	2.058e-01	1.177e-01	1.177e-01
1.00	0.95	2.057e-01	1.177e-01	1.177e-01
1.00	1.00	2.056e-01	1.177e-01	1.177e-01

ncall =   155

```
   t      r        m(r,t)
 0.00   0.00    1.000e+00
```

with $r = r_l$ corresponds to 0.5*(1+cos(pi*(r[ir]-rl)/(ru-rl)))
=0.5*(1+cos(pi*0))=1.

```
   t      r        m(r,t)
 0.00   1.00    0.000e+00
```

with $r = r_u$ corresponds to 0.5*(1+cos(pi*(r[ir]-rl)/(ru-rl)))
=0.5*(1+cos((pi*1))=0.
This verification is important since if the ICs are incorrect, the subsequent solution will be incorrect.

- The output is for $t = 0, 1$ as programmed in Listing 2.1.
- As in Table 2.1, the output is for $r = 0, 1/(21 - 1) = 0.05, \ldots, 1$ as programmed in Listing 2.1 ($3(nr) + 1 = 3(21) + 1 = 64$) values at each t. The $+1$ reflects the value of t at each of the 63 ODE solutions in matrix out from lsodes.
- The values of $m(r,t), c(r,t), a(r,t)$ are for the IC 0.5*(1+cos(pi*(r[ir] -rl)/(ru-rl))) (ncase=2), and reflect the cos and homogeneous Neumman BCs.
- $c(r,t)$ and $a(r,t)$ are the same since the MOL ODEs are the same for the ICs and parameter values in Listing 2.1, particularly tau=1 (see the programming for $\dfrac{\partial c}{\partial t}, \dfrac{\partial a}{\partial t}$ in pde1a of Listing 2.2).
- The computational effort is modest, ncall = 155, so lsodes calculates the solution efficiently.

The graphical output is in Figs. 2.2-1a, 2.2-1b, 2.2-2–2.2-3.

Figure 2.2-1a indicates that $m(r,t)$ starts from the IC in Listing 2.1 for ncase=2.

Figure 2.2-1b indicates the depletion of $m(r,t)$ starting from the IC in Listing 2.1 for ncase=2.

Figure 2.2-2 indicates that $c(r,t)$ results (is produced) from $m(r,t)$.

Figure 2.2-3 indicates that $a(r,t)$ results (is produced) from $m(r,t)$.

$m(r,t), c(r,t), a(r,t)$ have a long-term transient solution that can be studied by increasing tf (defined numerically in the main program of Listing 2.1). In particular, this transient reflects the chemotaxis term in eq. (1.3-1) (as discussed in subsequent chapters).

This completes the discussion of the main program and ODE/MOL routine in Listings 2.1 and 2.2.

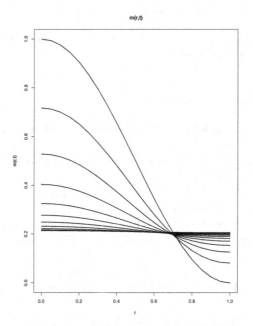

Figure 2.2-1a   Numerical solution $m(r,t)$ from eqs. (1.3-1, 1.3-4), 2D, `ncase=2`.

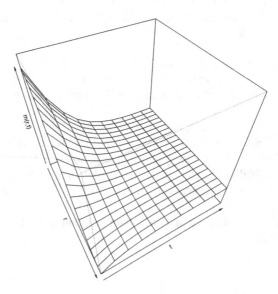

Figure 2.2-1b   Numerical solution $m(r,t)$ from eqs. (1.3-1, 1.3-4), 3D, `ncase=2`.

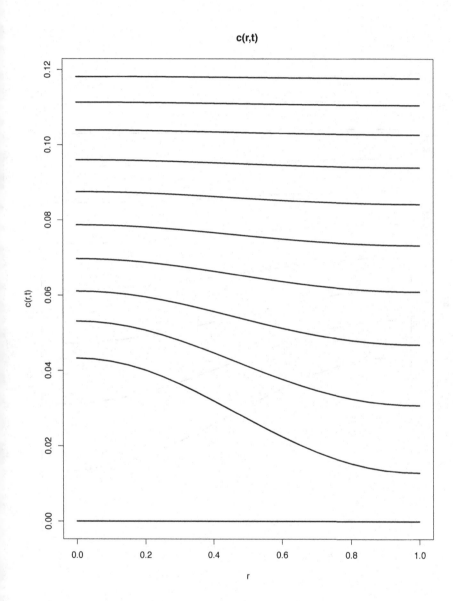

Figure 2.2-2 $c(r,t)$ from eqs. (1.3-2, 1.3-5), ncase=2.

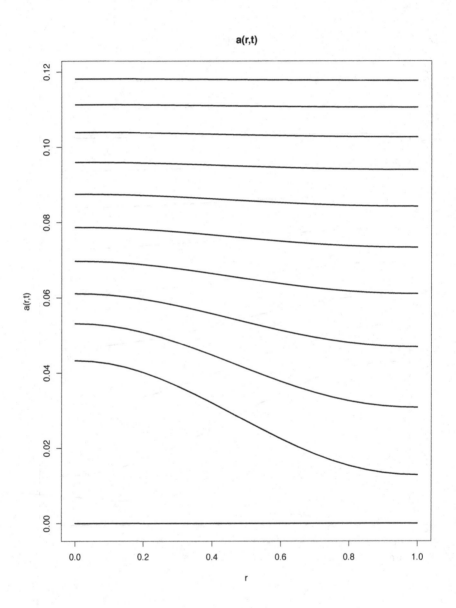

Figure 2.2-3 $a(r,t)$ from eqs. (1.3-3, 1.3-6), `ncase=2`.

Summary and conclusion

The chemotaxis PDE model that leads to inflammation from the macrophage cells is stated as eqs. (1.3-1–1.3-6) and (1.4-1–1.4-3). The ICs specified in the R main program discussed in Chapter 2 can produce a spectrum of solutions, as demonstrated for `ncase=1,2`. Generally, the pro-inflammatory and anti-inflamantory cytokines produced from the macrophage cells increase with t.

References

1. Giunta, M., Carmela Lombardo, and M. Sammartino (2021), Pattern Formation and Transition to Chaos in a Chemotaxis Model of Acute Inflammation. *SIAM Journal on Applied Dynamical Systems*, **20**, no. 4, pp. 1844–1881.

2. Soetaert, K., J. Cash, and F. Mazzia (2012), *Solving Differential Equations in R*, Springer-Verlag, Heidelberg, Germany.

3 Analysis of the Chemotaxis Model Time Derivatives

3.1 Introduction

The complex PDE model response in Table 2.2 and Figs. 2.2-1a, 2.2-1b to 2.2-3 (ncase=2) is determined by eqs. (1.3-1–1.3-6). Specifically, the derivatives $\frac{\partial m}{\partial t}, \frac{\partial c}{\partial t}, \frac{\partial a}{\partial t}$ of eqs. (1.3-1–1.3-6) define the solutions $m(r,t), c(r,t), a(r,t)$ [1]. Therefore, the focus of this chapter is the calculation and display of these derivatives.

3.2 Analysis of the chemotaxis model time derivatives

The following main program for eqs. (1.3-1–1.3-6) is again based on the use of the basic R system [2]. It is essentially the main program of Listing 2.1 for the ncase=2 IC and code added at the end for the calculation and display of the time derivatives of eqs. (1.3-1–1.3-6).

3.2.1 Main program

The main program for the analyis of the time derivatives is as follows:

```
#
# Time derivatives, PDE chemotaxis model
#
# Access ODE integrator
  library("deSolve");
#
# Access functions for numerical solution
  setwd("f:/inflammation/chap3");
  source("pde1a.R");
  source("dss004.R");
  source("dss044.R");
#
# Parameters
  D=0.25;
  rm=1;
  chi=1;
  alpha=1;
  beta=1;
```

DOI: 10.1201/9781003311201-3

```
  p=1;
  tau=1;
  m0=1;
#
# Grid (in r)
  nr=21;rl=0;ru=1
  r=seq(from=rl,to=ru,by=(ru-rl)/(nr-1));
#
# Independent variable for ODE integration
  t0=0;tf=1;nout=11;
  tout=seq(from=t0,to=tf,by=(tf-t0)/(nout-1));
#
# Initial condition
  u0=rep(0,3*nr);
    for(ir in 1:nr){
      u0[ir]     =0.5*(1+cos(pi*(r[ir]-rl)/(ru-rl)));
      u0[ir+nr] =0;
      u0[ir+2*nr]=0;
    }
  ncall=0;
#
# ODE integration
  out=lsodes(y=u0,times=tout,func=pde1a,
      sparsetype ="sparseint",rtol=1e-6,
      atol=1e-6,maxord=5);
  nrow(out)
  ncol(out)
#
# Arrays for plotting numerical solution
  m=matrix(0,nrow=nr,ncol=nout);
  c=matrix(0,nrow=nr,ncol=nout);
  a=matrix(0,nrow=nr,ncol=nout);
  for(it in 1:nout){
    for(ir in 1:nr){
      m[ir,it]=out[it,ir+1];
      c[ir,it]=out[it,ir+1+nr];
      a[ir,it]=out[it,ir+1+2*nr];
    }
  }
#
# Display numerical solution
  for(it in 1:nout){
    if((it==1)|(it==nout)){
      cat(sprintf("\n t   r   m(r,t)   c(r,t)   a(r,t)\n"));
      for(ir in 1:nr){
```

```
        cat(sprintf("%6.2f%6.2f%12.3e%12.3e%12.3e\n",
           tout[it],r[ir],m[ir,it],c[ir,it],a[ir,it])));
      }
    }
  }
#
# Calls to ODE routine
  cat(sprintf("\n\n ncall = %5d\n\n",ncall));
#
# Plot PDE solutions
#
# m
  par(mfrow=c(1,1));
  matplot(x=r,y=m,type="l",xlab="r",ylab="m(r,t)",
         xlim=c(rl,ru),lty=1,main="m(r,t)",lwd=2,
         col="black");
#
# c
  par(mfrow=c(1,1));
  matplot(x=r,y=c,type="l",xlab="r",ylab="c(r,t)",
         xlim=c(rl,ru),lty=1,main="c(r,t)",lwd=2,
         col="black");
#
# a
  par(mfrow=c(1,1));
  matplot(x=r,y=a,type="l",xlab="r",ylab="a(r,t)",
         xlim=c(rl,ru),lty=1,main="a(r,t)",lwd=2,
         col="black");
#
# mr,cr,ar
  mr=matrix(0,nrow=nr,ncol=nout);
  cr=matrix(0,nrow=nr,ncol=nout);
  ar=matrix(0,nrow=nr,ncol=nout);
  for(it in 1:nout){
    mr1=dss004(rl,ru,nr,m[,it]);
    cr1=dss004(rl,ru,nr,c[,it]);
    ar1=dss004(rl,ru,nr,a[,it]);
    mr[,it]=mr1;
    cr[,it]=cr1;
    ar[,it]=ar1;
  }
#
# BCs
  for(it in 1:nout){
    mr[1,it]=0;mr[nr,it]=0;
```

```
      cr[1,it]=0;cr[nr,it]=0;
      ar[1,it]=0;ar[nr,it]=0;
    }
#
# mrr,crr,arr
  mrr=matrix(0,nrow=nr,ncol=nout);
  crr=matrix(0,nrow=nr,ncol=nout);
  arr=matrix(0,nrow=nr,ncol=nout);
  nl=2;nu=2;
  for(it in 1:nout){
    mrr1=dss044(rl,ru,nr,m[,it],mr[,it],nl,nu);
    crr1=dss044(rl,ru,nr,c[,it],cr[,it],nl,nu);
    arr1=dss044(rl,ru,nr,a[,it],ar[,it],nl,nu);
    mrr[,it]=mrr1;
    crr[,it]=crr1;
    arr[,it]=arr1;
  }
#
# Functions of m,c,a
  func1=matrix(0,nrow=nr,ncol=nout);
  func2=matrix(0,nrow=nr,ncol=nout);
  for(it in 1:nout){
  for(ir in 1:nr){
    func1[ir,it]=chi*m[ir,it]/(1+alpha*c[ir,it])^2;
    func2[ir,it]=m[ir,it]/(1+beta*a[ir,it]^p);
  }
  }
  dfunc1=matrix(0,nrow=nr,ncol=nout);
  for(it in 1:nout){
   dfunc11=dss004(rl,ru,nr,func1[,it]);
   dfunc1[,it]=dfunc11;
  }
#
# PDEs
  mt=matrix(0,nrow=nr,ncol=nout);
  ct=matrix(0,nrow=nr,ncol=nout);
  at=matrix(0,nrow=nr,ncol=nout);
  for (it in 1:nout){
    for(ir in 1:nr){
    if(ir==1){
      mt[ir,it]=3*D*mrr[ir,it]-3*func1[ir,it]*crr[ir,it]+
        rm*m[ir,it]*c[ir,it]*(1-m[ir,it]);
      ct[ir,it]=3*crr[ir,it]+func2[ir,it]-c[ir,it];
      at[ir,it]=(1/tau)*(3*arr[ir,it]+func2[ir,it]-a[ir,it]);
    }
```

```
  if(ir==nr){
    mt[ir,it]=D*mrr[ir,it]-func1[ir,it]*crr[ir,it]+
      rm*m[ir,it]*c[ir,it]*(1-m[ir,it]);
    ct[ir,it]=crr[ir,it]+func2[ir,it]-c[ir,it];
    at[ir,it]=(1/tau)*(arr[ir,it]+func2[ir,it]-a[ir,it]);
    }
  if((ir>1)&(ir<nr)){
    mt[ir,it]=D*(mrr[ir,it]+(2/r[ir])*mr[ir,it])-
      dfunc1[ir,it]*cr[ir,it]-func1[ir,it]*crr[ir,it]-
        func1[ir,it]*(2/r[ir])*cr[ir,it]+
      rm*m[ir,it]*c[ir,it]*(1-m[ir,it]);
    ct[ir,it]=(crr[ir,it]+(2/r[ir])*cr[ir,it])+func2[ir,it]-
      c[ir,it];
    at[ir,it]=(1/tau)*(arr[ir,it]+(2/r[ir])*ar[ir,it]+
      func2[ir,it]-a[ir,it]);
    }
  }
 }
#
# Display t derivatives
  for(it in 1:nout){
    if((it==1)|(it==nout)){
    cat(sprintf("\n t   r   mt(r,t) ct(r,t) at(r,t)\n"));
    for(ir in 1:nr){
      cat(sprintf("%6.2f%6.2f%12.3e%12.3e%12.3e\n",
        tout[it],r[ir],mt[ir,it],ct[ir,it],at[ir,it]));
    }
   }
  }
#
# Plot PDE t derivatives
#
# mt
  par(mfrow=c(1,1));
  matplot(x=r,y=mt,type="l",xlab="r",ylab="mt(r,t)",
        xlim=c(rl,ru),lty=1,main="mt(r,t)",lwd=2,
        col="black");
#
# ct
  par(mfrow=c(1,1));
  matplot(x=r,y=ct,type="l",xlab="r",ylab="ct(r,t)",
        xlim=c(rl,ru),lty=1,main="ct(r,t)",lwd=2,
        col="black");
#
# at
```

```
par(mfrow=c(1,1));
matplot(x=r,y=at,type="l",xlab="r",ylab="at(r,t)",
        xlim=c(rl,ru),lty=1,main="at(r,t)",lwd=2,
        col="black");
```

Listing 3.1   Main program for eqs. (1.3-1–1.3-6)

We can note the following details about Listing 3.1 (in addition to the details for
Listing 2.1).

- The IC is again (as for ncase=2 in Listing 2.1) a half cos wave.

```
#
# Initial condition
  u0=rep(0,3*nr);
    for(ir in 1:nr){
      u0[ir]       =0.5*(1+cos(pi*(r[ir]-rl)/(ru-rl)));
      u0[ir+nr]    =0;
      u0[ir+2*nr]=0;
    }
```

- At the end of Listing 3.1, the time derivatives of eqs. (1.3-1–1.3-6) are calculated and displayed, starting with the right hand side (RHS) terms of eqs.
  (1.3-1–1.3-6). The derivatives $\frac{\partial m}{\partial r}, \frac{\partial c}{\partial r}, \frac{\partial a}{\partial r}$, are calculated with dss004
  and placed in the 2D arrays mr,cr,ar.

```
#
# mr,cr,ar
  mr=matrix(0,nrow=nr,ncol=nout);
  cr=matrix(0,nrow=nr,ncol=nout);
  ar=matrix(0,nrow=nr,ncol=nout);
  for(it in 1:nout){
    mr1=dss004(rl,ru,nr,m[,it]);
    cr1=dss004(rl,ru,nr,c[,it]);
    ar1=dss004(rl,ru,nr,a[,it]);
    mr[,it]=mr1;
    cr[,it]=cr1;
    ar[,it]=ar1;
  }
```

Since dss004 returns derivatives as 1D vectors, they are placed in
mr1,cr1,ar1, and then transferred to 2D arrays mr,cr,ar for subsequent
calculations. The input vectors to dss004, are 1D, m[,it],c[,it],a[,it]
where a comma is used for the nr=21 values of r. The particular value of t
is specified with the index it.
The calculation of the first derivatives in r is the same as in pde1a of Listing
2.2, but with a second subscript it added for t.

- BCs (1.4-1–1.4-3) (homogeneous Neumann, no flux conditions) are programmed.

```
#
# BCs
  for(it in 1:nout){
    mr[1,it]=0;mr[nr,it]=0;
    cr[1,it]=0;cr[nr,it]=0;
    ar[1,it]=0;ar[nr,it]=0;
  }
```

Subscripts 1, nr correspond to $r = r_l = 0$, $r = r_u = 1$, respectively.

- The second derivatives $\dfrac{\partial^2 m}{\partial r^2}$, $\dfrac{\partial^2 c}{\partial r^2}$, $\dfrac{\partial^2 a}{\partial r^2}$, are calculated with dss044 and placed in the 2D arrays mrr,crr,arr (in analogy with the programming of first spatial derivatives discussed previously).

```
#
# mrr,crr,arr
  mrr=matrix(0,nrow=nr,ncol=nout);
  crr=matrix(0,nrow=nr,ncol=nout);
  arr=matrix(0,nrow=nr,ncol=nout);
  nl=2;nu=2;
  for(it in 1:nout){
    mrr1=dss044(rl,ru,nr,m[,it],mr[,it],nl,nu);
    crr1=dss044(rl,ru,nr,c[,it],cr[,it],nl,nu);
    arr1=dss044(rl,ru,nr,a[,it],ar[,it],nl,nu);
    mrr[,it]=mrr1;
    crr[,it]=crr1;
    arr[,it]=arr1;
  }
```

nl=2,nu=2 specify Neumann BCs.

- Functions

$$\chi\frac{m}{(1+\alpha c)^2} = \text{func1}, \quad \frac{m}{1+\beta a^p} = \text{func2}, \quad \frac{\partial\left(\chi\dfrac{m}{(1+\alpha c)^2}\right)}{\partial r} = \text{dfunc1 are}$$

programmed (for use in eqs. (1.3-1–1.3-6)).

```
#
# Functions of m,c,a
  func1=matrix(0,nrow=nr,ncol=nout);
  func2=matrix(0,nrow=nr,ncol=nout);
  for(it in 1:nout){
  for(ir in 1:nr){
    func1[ir,it]=chi*m[ir,it]/(1+alpha*c[ir,it])^2;
    func2[ir,it]=m[ir,it]/(1+beta*a[ir,it]^p);
  }
```

```
}
dfunc1=matrix(0,nrow=nr,ncol=nout);
for(it in 1:nout){
 dfunc11=dss004(rl,ru,nr,func1[,it]);
 dfunc1[,it]=dfunc11;
}
```

- The derivatives (left hand side (LHS) terms), of eqs. (1.3-1–1.3-6)) $\frac{\partial m}{\partial t}$, $\frac{\partial c}{\partial t}$, $\frac{\partial a}{\partial t}$ are computed and placed in arrays mt,ct,at.

```
#
# PDEs
 mt=matrix(0,nrow=nr,ncol=nout);
 ct=matrix(0,nrow=nr,ncol=nout);
 at=matrix(0,nrow=nr,ncol=nout);
 for (it in 1:nout){
   for(ir in 1:nr){
   if(ir==1){
     mt[ir,it]=3*D*mrr[ir,it]-3*func1[ir,it]*crr[ir,it]+
       rm*m[ir,it]*c[ir,it]*(1-m[ir,it]);
     ct[ir,it]=3*crr[ir,it]+func2[ir,it]-c[ir,it];
     at[ir,it]=(1/tau)*(3*arr[ir,it]+func2[ir,it]
      -a[ir,it]);
   }
  if(ir==nr){
     mt[ir,it]=D*mrr[ir,it]-func1[ir,it]*crr[ir,it]+
       rm*m[ir,it]*c[ir,it]*(1-m[ir,it]);
     ct[ir,it]=crr[ir,it]+func2[ir,it]-c[ir,it];
     at[ir,it]=(1/tau)*(arr[ir,it]+func2[ir,it]-a[ir,it]);
   }
   if((ir>1)&(ir<nr)){
     mt[ir,it]=D*(mrr[ir,it]+(2/r[ir])*mr[ir,it])-
       dfunc1[ir,it]*cr[ir,it]-func1[ir,it]*crr[ir,it]-
        func1[ir,it]*(2/r[ir])*cr[ir,it]+
       rm*m[ir,it]*c[ir,it]*(1-m[ir,it]);
     ct[ir,it]=(crr[ir,it]+(2/r[ir])*cr[ir,it])
      +func2[ir,it]-
      c[ir,it];
     at[ir,it]=(1/tau)*(arr[ir,it]+(2/r[ir])*ar[ir,it]+
       func2[ir,it]-a[ir,it]);
   }
  }
 }
```

This coding follows from the calculation of the t derivatives in pde1a of Listing 2.2 with a second index, it, added for t.

- The derivatives $\dfrac{\partial m}{\partial t}, \dfrac{\partial c}{\partial t}, \dfrac{\partial a}{\partial t}$ are plotted in 2D with the matplot utility.

```
#
# Plot PDE t derivatives
#
# mt
  par(mfrow=c(1,1));
  matplot(x=r,y=mt,type="l",xlab="r",ylab="mt(r,t)",
          xlim=c(rl,ru),lty=1,main="mt(r,t)",lwd=2,
          col="black");
#
# ct
  par(mfrow=c(1,1));
  matplot(x=r,y=ct,type="l",xlab="r",ylab="ct(r,t)",
          xlim=c(rl,ru),lty=1,main="ct(r,t)",lwd=2,
          col="black");
#
# at
  par(mfrow=c(1,1));
  matplot(x=r,y=at,type="l",xlab="r",ylab="at(r,t)",
          xlim=c(rl,ru),lty=1,main="at(r,t)",lwd=2,
          col="black");
```

This completes the discussion of the main program in Listing 3.1.

3.2.2 ODE/MOL routine

The ODE/MOL routine pde1a is the same as in Listing 2.2.

3.2.3 Numerical, graphical output

Abbreviated output from Listings 3.1 and 2.2 follows (see Table 3.1):
We can note the following details about this output.

- $m(r,t), c(r,t), a(r,t)$ are not included since they are the same as in Table 2.2.
- The output is for $t = t_0 = 0, t = t_f = 1$ as programmed in Listing 3.1.
- The t derivatives, $\dfrac{\partial m}{\partial t} = $ mt, $\dfrac{\partial c}{\partial t} = $ ct, $\dfrac{\partial a}{\partial t} = $ at, have substantial values above and below zero.

The graphical output is in Figs. 3.1-1–3.1-3.

Figure 3.1-1 indicates that $\dfrac{\partial m(r,t)}{\partial t}$ has a zero derivative (slope) at $r = r_l = 0$, $r = r_u = 1$ as expected (from homogeneous Neumann BCs (1.4-1–1.4-3)), For example, at $t = 0$,

$$\frac{d\left(0.5(1 + \cos(\pi(r - r_l)/(r_u - r_l)))\right)}{dr} = -0.5\pi/(r_u - r_l)\sin(\pi(r - r_l)/(r_u - r_l)) = 0;$$

$$r = r_l = 0, r = r_u = 1$$

Table 3.1

Numerical t Derivatives for eqs. (1.3-1–1.3-6)

t	r	mt(r,t)	ct(r,t)	at(r,t)
0.00	0.00	-3.701e+00	1.000e+00	1.000e+00
0.00	0.05	-3.676e+00	9.938e-01	9.938e-01
0.00	0.10	-3.600e+00	9.755e-01	9.755e-01
0.00	0.15	-3.476e+00	9.455e-01	9.455e-01
0.00	0.20	-3.306e+00	9.045e-01	9.045e-01
.			.	
.			.	
.			.	

Output for r=0.25 to 0.75 deleted

.			.	
.			.	
.			.	
0.00	0.80	4.210e-01	9.549e-02	9.549e-02
0.00	0.85	6.798e-01	5.450e-02	5.450e-02
0.00	0.90	9.036e-01	2.447e-02	2.447e-02
0.00	0.95	1.089e+00	6.156e-03	6.156e-03
0.00	1.00	1.234e+00	0.000e+00	0.000e+00

t	r	mt(r,t)	ct(r,t)	at(r,t)
1.00	0.00	-1.488e-02	6.569e-02	6.569e-02
1.00	0.05	-1.459e-02	6.571e-02	6.571e-02
1.00	0.10	-1.374e-02	6.575e-02	6.575e-02
1.00	0.15	-1.234e-02	6.583e-02	6.583e-02
1.00	0.20	-1.044e-02	6.593e-02	6.593e-02
.			.	
.			.	
.			.	

Output for r=0.25 to 0.75 deleted

.			.	
.			.	
.			.	
1.00	0.80	2.351e-02	6.780e-02	6.780e-02
1.00	0.85	2.495e-02	6.788e-02	6.788e-02
1.00	0.90	2.597e-02	6.793e-02	6.793e-02
1.00	0.95	2.656e-02	6.797e-02	6.797e-02
1.00	1.00	2.676e-02	6.798e-02	6.798e-02

mt(r,t)

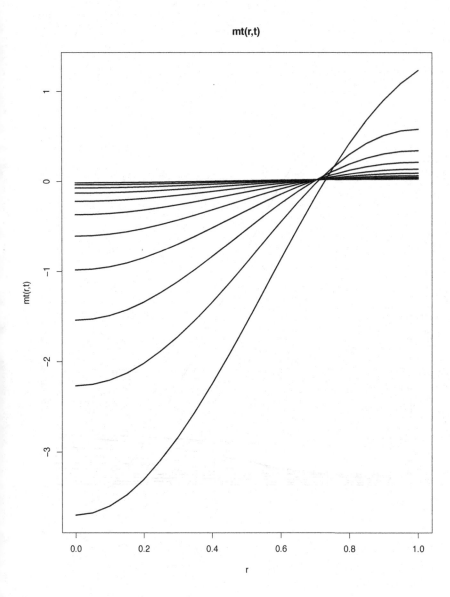

Figure 3.1-1 Numerical derivative $\dfrac{\partial m(r,t)}{\partial t}$.

Also, Figure 3.1-1 and Table 3.1 indicate that $m(r,t)$ may approach a small, non-zero equilibrium (steady state) solution ($\dfrac{\partial m(r,t)}{\partial t}$ does not approach zero). This conclusion could be studied by increasing $t = t_f$ (set in Listing 3.1). This is left as an exercise.

ct(r,t)

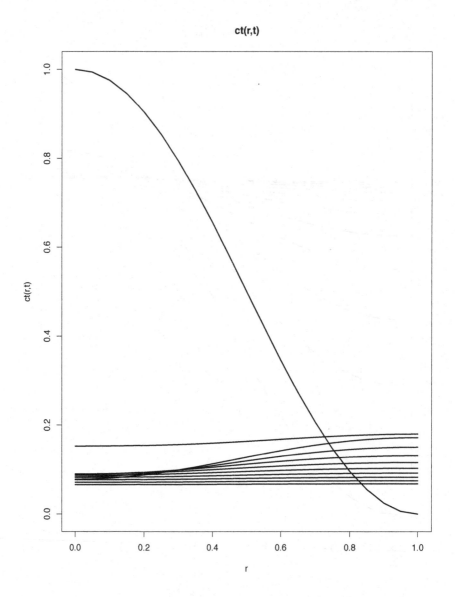

Figure 3.1-2   Numerical derivative $\dfrac{\partial c(r,t)}{\partial t}$.

Figure 3.1-2 indicates that $c(r,t)$ appears to reach a non-zero equilibrium (steady state) solution ($\dfrac{\partial c(r,t)}{\partial t}$ does not approach zero).

Figure 3.1-3 indicates that $a(r,t)$ appears to reach a non-zero equilibrium (steady state) solution ($\dfrac{\partial a(r,t)}{\partial t}$ does not approach zero).

at(r,t)

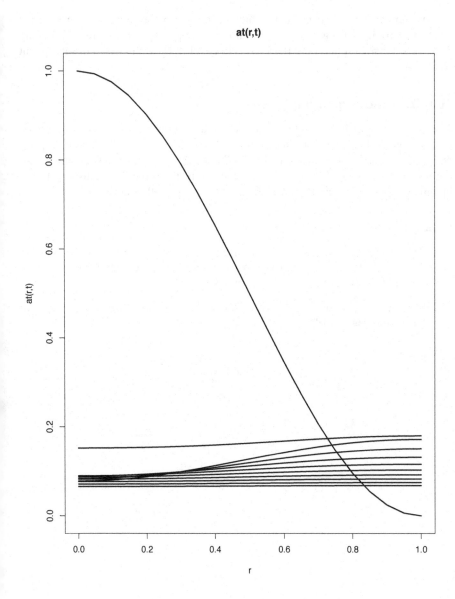

Figure 3.1-3 Numerical derivative $\dfrac{\partial a(r,t)}{\partial t}$.

Also,

$$\frac{\partial a(r,t)}{\partial t} = \frac{\partial c(r,t)}{\partial t}$$

since the RHSs of eqs. (1.3-2), (1.3-5), (1.3-3), and (1.3-6) are the same for the parameters in Listing 3.1 and the programming in Listing 2.2.

In summary, $m(r,t), c(r,t), a(r,t)$ have a long-term transient solution that can be studied by increasing `tf` (defined numerically in the main program of Listing 3.1). This transient reflects the RHS terms in eq. (1.3-1) (as discussed in subsequent chapters).

Summary and conclusion

The chemotaxis PDE model that leads to inflammation from the depletion of the macrophage cells is stated as eqs. (1.3-1–1.3-6) and (1.4-1–1.4-3). The ICs specified in the R main program discussed in Chapter 2 can produce a spectrum of solutions, as demonstrated for `ncase=1,2`. Generally, the pro-inflammatory and anti-inflamantory cytokines produced from the macrophage cells increase with t.

References

1. Giunta, M., Carmela Lombardo, and M. Sammartino (2021), Pattern Formation and Transition to Chaos in a Chemotaxis Model of Acute Inflammation. *SIAM Journal on Applied Dynamical Systems*, **20**, no. 4, pp. 1844–1881.

2. Soetaert, K., J. Cash, and F. Mazzia (2012), *Solving Differential Equations in R*, Springer-Verlag, Heidelberg, Germany.

4 Analysis of the Chemotaxis PDE Model Terms

4.1 Introduction

The numerical solutions $m(r,t), c(r,t), a(r,t)$ to eqs. (1.3-1–1.3-6) are computed in Chapter 2 by the method of lines (MOL) (Listings 2.1 and 2.2). The features of these solutions result from the derivatives $\dfrac{\partial m}{\partial t}, \dfrac{\partial c}{\partial t}, \dfrac{\partial a}{\partial t}$ computed in Chapter 3 (Listings 3.1 and 2.2).

The terms in the PDEs are computed and displayed in this chapter as further explanation of the features of the PDE solutions [1,2]. The R routines for the PDE terms consist of a main program that follows and the ODE/MOL routine of Listing 2.2.

4.2 R routines for PDE RHS, LHS terms

The main program with the calculation of the PDE terms follows.

4.2.1 Main program

The main program is the same as in Listing 3.1 with additional programming at the end for the calculation and display of the PDE RHS, LHS terms of eqs. (1.3-1–1.3-6).

```
#
#  RHS, LHS PDE terms, chemotaxis model
#
# Access ODE integrator
  library("deSolve");
#
# Access functions for numerical solution
  setwd("f:/inflammation/chap4");
  source("pde1a.R");
  source("dss004.R");
  source("dss044.R");
#
# Parameters
  D=0.25;
  rm=1;
  chi=1;
  alpha=1;
  beta=1;
```

DOI: 10.1201/9781003311201-4

```
  p=1;
  tau=1;
  m0=1;
#
# Grid (in r)
  nr=21;rl=0;ru=1
  r=seq(from=rl,to=ru,by=(ru-rl)/(nr-1));
#
# Independent variable for ODE integration
  t0=0;tf=1;nout=11;
  tout=seq(from=t0,to=tf,by=(tf-t0)/(nout-1));
#
# Initial condition
  u0=rep(0,3*nr);
    for(ir in 1:nr){
      u0[ir]      =0.5*(1+cos(pi*(r[ir]-rl)/(ru-rl)));
      u0[ir+nr]  =0;
      u0[ir+2*nr]=0;
    }
  ncall=0;
#
# ODE integration
  out=lsodes(y=u0,times=tout,func=pde1a,
      sparsetype ="sparseint",rtol=1e-6,
      atol=1e-6,maxord=5);
  nrow(out)
  ncol(out)
#
# Arrays for plotting numerical solution
  m=matrix(0,nrow=nr,ncol=nout);
  c=matrix(0,nrow=nr,ncol=nout);
  a=matrix(0,nrow=nr,ncol=nout);
  for(it in 1:nout){
    for(ir in 1:nr){
      m[ir,it]=out[it,ir+1];
      c[ir,it]=out[it,ir+1+nr];
      a[ir,it]=out[it,ir+1+2*nr];
    }
  }
#
# Display numerical solution
  for(it in 1:nout){
    if((it==1)|(it==nout)){
      cat(sprintf("\n t    r    m(r,t)   c(r,t)   a(r,t)\n"));
      for(ir in 1:nr){
```

```
        cat(sprintf("%6.2f%6.2f%12.3e%12.3e%12.3e\n",
          tout[it],r[ir],m[ir,it],c[ir,it],a[ir,it]));
      }
    }
  }
#
# Calls to ODE routine
  cat(sprintf("\n\n ncall = %5d\n\n",ncall));
#
# Plot PDE solutions
#
# m
  par(mfrow=c(1,1));
  matplot(x=r,y=m,type="l",xlab="r",ylab="m(r,t)",
          xlim=c(rl,ru),lty=1,main="m(r,t)",lwd=2,
          col="black");
#
# c
  par(mfrow=c(1,1));
  matplot(x=r,y=c,type="l",xlab="r",ylab="c(r,t)",
          xlim=c(rl,ru),lty=1,main="c(r,t)",lwd=2,
          col="black");
#
# a
  par(mfrow=c(1,1));
  matplot(x=r,y=a,type="l",xlab="r",ylab="a(r,t)",
          xlim=c(rl,ru),lty=1,main="a(r,t)",lwd=2,
          col="black");
#
# mr,cr,ar
  mr=matrix(0,nrow=nr,ncol=nout);
  cr=matrix(0,nrow=nr,ncol=nout);
  ar=matrix(0,nrow=nr,ncol=nout);
  for(it in 1:nout){
    mr1=dss004(rl,ru,nr,m[,it]);
    cr1=dss004(rl,ru,nr,c[,it]);
    ar1=dss004(rl,ru,nr,a[,it]);
    mr[,it]=mr1;
    cr[,it]=cr1;
    ar[,it]=ar1;
  }
#
# BCs
  for(it in 1:nout){
    mr[1,it]=0;mr[nr,it]=0;
```

```
      cr[1,it]=0;cr[nr,it]=0;
      ar[1,it]=0;ar[nr,it]=0;
    }
#
# mrr,crr,arr
  mrr=matrix(0,nrow=nr,ncol=nout);
  crr=matrix(0,nrow=nr,ncol=nout);
  arr=matrix(0,nrow=nr,ncol=nout);
  nl=2;nu=2;
  for(it in 1:nout){
    mrr1=dss044(rl,ru,nr,m[,it],mr[,it],nl,nu);
    crr1=dss044(rl,ru,nr,c[,it],cr[,it],nl,nu);
    arr1=dss044(rl,ru,nr,a[,it],ar[,it],nl,nu);
    mrr[,it]=mrr1;
    crr[,it]=crr1;
    arr[,it]=arr1;
  }
#
# Functions of m,c,a
  func1=matrix(0,nrow=nr,ncol=nout);
  func2=matrix(0,nrow=nr,ncol=nout);
  for(it in 1:nout){
  for(ir in 1:nr){
    func1[ir,it]=chi*m[ir,it]/(1+alpha*c[ir,it])^2;
    func2[ir,it]=m[ir,it]/(1+beta*a[ir,it]^p);
  }
  }
  dfunc1=matrix(0,nrow=nr,ncol=nout);
  for(it in 1:nout){
   dfunc11=dss004(rl,ru,nr,func1[,it]);
   dfunc1[,it]=dfunc11;
  }
#
# PDEs
  mt=matrix(0,nrow=nr,ncol=nout);
  ct=matrix(0,nrow=nr,ncol=nout);
  at=matrix(0,nrow=nr,ncol=nout);
  for (it in 1:nout){
    for(ir in 1:nr){
    if(ir==1){
      mt[ir,it]=3*D*mrr[ir,it]-3*func1[ir,it]*crr[ir,it]+
        rm*m[ir,it]*c[ir,it]*(1-m[ir,it]);
      ct[ir,it]=3*crr[ir,it]+func2[ir,it]-c[ir,it];
      at[ir,it]=(1/tau)*(3*arr[ir,it]+func2[ir,it]-a[ir,it]);
    }
```

```
  if(ir==nr){
    mt[ir,it]=D*mrr[ir,it]-func1[ir,it]*crr[ir,it]+
      rm*m[ir,it]*c[ir,it]*(1-m[ir,it]);
    ct[ir,it]=crr[ir,it]+func2[ir,it]-c[ir,it];
    at[ir,it]=(1/tau)*(arr[ir,it]+func2[ir,it]-a[ir,it]);
  }
  if((ir>1)&(ir<nr)){
    mt[ir,it]=D*(mrr[ir,it]+(2/r[ir])*mr[ir,it])-
      dfunc1[ir,it]*cr[ir,it]-func1[ir,it]*crr[ir,it]-
        func1[ir,it]*(2/r[ir])*cr[ir,it]+
      rm*m[ir,it]*c[ir,it]*(1-m[ir,it]);
    ct[ir,it]=(crr[ir,it]+(2/r[ir])*cr[ir,it])+func2[ir,it]-
      c[ir,it];
    at[ir,it]=(1/tau)*(arr[ir,it]+(2/r[ir])*ar[ir,it]+
      func2[ir,it]-a[ir,it]);
  }
}
}
#
# Display t derivatives
  for(it in 1:nout){
    if((it==1)|(it==nout)){
      cat(sprintf("\n t   r    mt(r,t) ct(r,t) at(r,t)\n"));
      for(ir in 1:nr){
        cat(sprintf("%6.2f%6.2f%12.3e%12.3e%12.3e\n",
          tout[it],r[ir],mt[ir,it],ct[ir,it],at[ir,it]));
      }
    }
  }
}
#
# Plot PDE t derivatives
#
# mt
  par(mfrow=c(1,1));
  matplot(x=r,y=mt,type="l",xlab="r",ylab="mt(r,t)",
          xlim=c(rl,ru),lty=1,main="mt(r,t)",lwd=2,
          col="black");
#
# ct
  par(mfrow=c(1,1));
  matplot(x=r,y=ct,type="l",xlab="r",ylab="ct(r,t)",
          xlim=c(rl,ru),lty=1,main="ct(r,t)",lwd=2,
          col="black");
#
# at
```

```
  par(mfrow=c(1,1));
  matplot(x=r,y=at,type="l",xlab="r",ylab="at(r,t)",
          xlim=c(rl,ru),lty=1,main="at(r,t)",lwd=2,
          col="black");
#
# PDE RHS terms
#
# m(r,t)
  term11=matrix(0,nrow=nr,ncol=nout);
  term12=matrix(0,nrow=nr,ncol=nout);
  term13=matrix(0,nrow=nr,ncol=nout);
  term14=matrix(0,nrow=nr,ncol=nout);
  term21=matrix(0,nrow=nr,ncol=nout);
  term22=matrix(0,nrow=nr,ncol=nout);
  term23=matrix(0,nrow=nr,ncol=nout);
  term24=matrix(0,nrow=nr,ncol=nout);
  term31=matrix(0,nrow=nr,ncol=nout);
  term32=matrix(0,nrow=nr,ncol=nout);
  term33=matrix(0,nrow=nr,ncol=nout);
  term34=matrix(0,nrow=nr,ncol=nout);
  for(it in 1:nout){
  for(ir in 1:nr){
    if(ir==1){term11[ir,it]=3*D*mrr[ir,it];
               term12[ir,it]=-3*func1[ir,it]*crr[ir,it];
               term13[ir,it]=rm*m[ir,it]*c[ir,it]*(1-m[ir,it]);
               term14[ir,it]=term11[ir,it]+term12[ir,it]
                 +term13[ir,it];
               term21[ir,it]=3*crr[ir,it];
               term22[ir,it]=func2[ir,it];
               term23[ir,it]=-c[ir,it];
               term24[ir,it]=term21[ir,it]+term22[ir,it]
                 +term23[ir,it];
               term31[ir,it]=(1/tau)*3*arr[ir,it];
               term32[ir,it]=(1/tau)*func2[ir,it];
               term33[ir,it]=-(1/tau)*a[ir,it];
               term34[ir,it]=term31[ir,it]+term32[ir,it]
                 +term33[ir,it];}
    if(ir==nr){term11[ir,it]=D*mrr[ir,it];
               term12[ir,it]=-func1[ir]*crr[ir];
               term13[ir,it]=rm*m[ir,it]*c[ir,it]*(1-m[ir,it]);
               term14[ir,it]=term11[ir,it]+term12[ir,it]
                 +term13[ir,it];
               term21[ir,it]=crr[ir,it];
               term22[ir,it]=func2[ir,it];
               term23[ir,it]=-c[ir,it];
```

```
                 term24[ir,it]=term21[ir,it]+term22[ir,it]
                   +term23[ir,it];
                 term31[ir,it]=(1/tau)*arr[ir,it];
                 term32[ir,it]=(1/tau)*func2[ir,it];
                 term33[ir,it]=-(1/tau)*a[ir,it];
                 term34[ir,it]=term31[ir,it]+term32[ir,it]
                   +term33[ir,it];}
     if((ir>1)&(ir<nr)){term11[ir,it]=D*(mrr[ir,it]+(2/r[ir])
                        *mr[ir,it]);
                 term12[ir,it]=dfunc1[ir,it]*cr[ir,it]-
                 func1[ir,it]*crr[ir,it]-
                 func1[ir,it]*(2/r[ir])*cr[ir,it];
                 term13[ir,it]=rm*m[ir,it]*c[ir,it]
                  *(1-m[ir,it]);
                 term14[ir,it]=term11[ir,it]+term12[ir,it]+
                              term13[ir,it];
                 term21[ir,it]=(crr[ir,it]+(2/r[ir])
                  *cr[ir,it]);
                 term22[ir,it]=func2[ir,it];
                 term23[ir,it]=-c[ir,it];
                 term24[ir,it]=term21[ir,it]+term22[ir,it]
                              +term23[ir,it];
                 term31[ir,it]=(1/tau)*(arr[ir,it]+(2/r[ir])
                            *ar[ir,it])
                 term32[ir,it]=(1/tau)*func2[ir,it];
                 term33[ir,it]=-(1/tau)*a[ir,it];
                 term34[ir,it]=term31[ir,it]+term32[ir,it]
                              +term33[ir,it];}
  }
  }
#
# Plot PDE RHS, LHS terms
#
# term11
  par(mfrow=c(2,2));
  matplot(x=r,y=term11,type="l",xlab="r",ylab="term11(r,t)",
          xlim=c(rl,ru),lty=1,main="term11(r,t)",lwd=2,
          col="black",ylim=c(-5,5));
#
# term12
  matplot(x=r,y=term12,type="l",xlab="r",ylab="term12(r,t)",
          xlim=c(rl,ru),lty=1,main="term12(r,t)",lwd=2,
          col="black",ylim=c(-5,5));
#
# term13
```

```
  matplot(x=r,y=term13,type="l",xlab="r",ylab="term13(r,t)",
        xlim=c(rl,ru),lty=1,main="term13(r,t)",lwd=2,
        col="black",ylim=c(-5,5));
#
# term14
  matplot(x=r,y=term14,type="l",xlab="r",ylab="term14(r,t)",
        xlim=c(rl,ru),lty=1,main="term14(r,t)",lwd=2,
        col="black",ylim=c(-5,5));
#
# term21
  par(mfrow=c(2,2));
  matplot(x=r,y=term21,type="l",xlab="r",ylab="term21(r,t)",
        xlim=c(rl,ru),lty=1,main="term21(r,t)",lwd=2,
        col="black",ylim=c(-1,1));
#
# term22
  matplot(x=r,y=term22,type="l",xlab="r",ylab="term22(r,t)",
        xlim=c(rl,ru),lty=1,main="term22(r,t)",lwd=2,
        col="black",ylim=c(-1,1));
#
# term23
  matplot(x=r,y=term23,type="l",xlab="r",ylab="term23(r,t)",
        xlim=c(rl,ru),lty=1,main="term23(r,t)",lwd=2,
        col="black",ylim=c(-1,1));
#
# term24
  matplot(x=r,y=term24,type="l",xlab="r",ylab="term24(r,t)",
        xlim=c(rl,ru),lty=1,main="term24(r,t)",lwd=2,
        col="black",ylim=c(-1,1));
#
# term31
  par(mfrow=c(2,2));
  matplot(x=r,y=term31,type="l",xlab="r",ylab="term31(r,t)",
        xlim=c(rl,ru),lty=1,main="term31(r,t)",lwd=2,
        col="black",ylim=c(-1,1));
#
# term32
  matplot(x=r,y=term32,type="l",xlab="r",ylab="term32(r,t)",
        xlim=c(rl,ru),lty=1,main="term32(r,t)",lwd=2,
        col="black",ylim=c(-1,1));
#
# term33
  matplot(x=r,y=term33,type="l",xlab="r",ylab="term33(r,t)",
        xlim=c(rl,ru),lty=1,main="term33(r,t)",lwd=2,
        col="black",ylim=c(-1,1));
```

```
#
# term34
  matplot(x=r,y=term34,type="l",xlab="r",ylab="term34(r,t)",
          xlim=c(rl,ru),lty=1,main="term34(r,t)",lwd=2,
          col="black",ylim=c(-1,1));
```
Listing 4.1   Main program for eqs. (1.3-1–1.3-6) with RHS, LHS term calculations

We can note the following details about Listing 4.1 (in addition to the details for Listings 2.1 and 3.1).

- ICs for eqs. (1.3-1) and (1.3-4) is again a half cosine wave.

```
#
# Initial condition
  u0=rep(0,3*nr);
    for(ir in 1:nr){
      u0[ir]      =0.5*(1+cos(pi*(r[ir]-rl)/(ru-rl)));
      u0[ir+nr]   =0;
      u0[ir+2*nr]=0;
    }
```

- At the end of Listing 3.1, the calculation and display of the PDE RHS, LHS terms is added, starting with arrays for the terms.

```
#
# PDE LHS, RHS terms
#
# m(r,t)
  term11=matrix(0,nrow=nr,ncol=nout);
  term12=matrix(0,nrow=nr,ncol=nout);
  term13=matrix(0,nrow=nr,ncol=nout);
  term14=matrix(0,nrow=nr,ncol=nout);
  term21=matrix(0,nrow=nr,ncol=nout);
  term22=matrix(0,nrow=nr,ncol=nout);
  term23=matrix(0,nrow=nr,ncol=nout);
  term24=matrix(0,nrow=nr,ncol=nout);
  term31=matrix(0,nrow=nr,ncol=nout);
  term32=matrix(0,nrow=nr,ncol=nout);
  term33=matrix(0,nrow=nr,ncol=nout);
  term34=matrix(0,nrow=nr,ncol=nout);
```

The naming of the terms follows `termij`, where `i` and `j` are the equation number and term number, respectively. For example, `term11` is the diffusion (first) term in (1) eq. (1.3-1) $(r > r_l)$, $D\left(\dfrac{\partial^2 m}{\partial r^2} + \dfrac{2}{r}\dfrac{\partial m}{\partial r}\right)$ or (2) eq. (1.3-4) $(r = r_l = 0)$, $3D\dfrac{\partial^2 m}{\partial r^2}$.

- The terms are computed as a function of the independent variables, r,t, within two fors. For the first PDE at $r = r_l = 0$ (ir=1), eq. (1.3-4), the RHS terms are term11, term12, term13, and the LHS term is term14 which is the time derivative $\dfrac{\partial m}{\partial t}$ that can be compared with the time derivative computed in Chapter 3.

```
for(it in 1:nout){
for(ir in 1:nr){
  if(ir==1){term11[ir,it]=3*D*mrr[ir,it];
            term12[ir,it]=-3*func1[ir,it]*crr[ir,it];
            term13[ir,it]=rm*m[ir,it]*c[ir,it]
            *(1-m[ir,it]);
            term14[ir,it]=term11[ir,it]+term12[ir,it]
                          +term13[ir,it];
```

This code includes BC (1.4-1).
- For the second PDE at $r = r_l = 0$ (ir=1), eq. (1.3-5), the RHS terms are term21, term22, term23, and the LHS term is term24 which is the time derivative $\dfrac{\partial c}{\partial t}$.

```
            term21[ir,it]=3*crr[ir,it];
            term22[ir,it]=func2[ir,it];
            term23[ir,it]=-c[ir,it];
            term24[ir,it]=term21[ir,it]+term22[ir,it]
                          +term23[ir,it];
```

This code includes BC (1.4-2).
- For the third PDE at $r = r_l = 0$ (ir=1), eq. (1.3-6), the RHS terms are term31, term32, term33, and the LHS term is term34 which is the time derivative $\dfrac{\partial a}{\partial t}$.

```
            term31[ir,it]=(1/tau)*3*arr[ir,it];
            term32[ir,it]=(1/tau)*func2[ir,it];
            term33[ir,it]=-(1/tau)*a[ir,it];
            term34[ir,it]=term31[ir,it]+term32[ir,it]
                          +term33[ir,it];}
```

This code includes BC (1.4-3).
- A similar analysis applies to the three PDEs at $r = r_u = 1$ (ir=nr), with application of BCs (1.4-1–1.4-3).

```
  if(ir==nr){term11[ir,it]=D*mrr[ir,it];
            term12[ir,it]=-func1[ir]*crr[ir];
            term13[ir,it]=rm*m[ir,it]*c[ir,it]
            *(1-m[ir,it]);
```

```
                    term14[ir,it]=term11[ir,it]+term12[ir,it]
                              +term13[ir,it];
                    term21[ir,it]=crr[ir,it];
                    term22[ir,it]=func2[ir,it];
                    term23[ir,it]=-c[ir,it];
                    term24[ir,it]=term21[ir,it]+term22[ir,it]
                              +term23[ir,it];
                    term31[ir,it]=(1/tau)*arr[ir,it];
                    term32[ir,it]=(1/tau)*func2[ir,it];
                    term33[ir,it]=-(1/tau)*a[ir,it];
                    term34[ir,it]=term31[ir,it]+term32[ir,it]
                              +term33[ir,it];}
```

- A similar analysis applies to the three PDEs at $r_l = 0 \le r \le r_u = 1$ ((ir>1)&(ir<nr)).

```
        if((ir>1)&(ir<nr)){term11[ir,it]=D*(mrr[ir,it]
                          +(2/r[ir])*mr[ir,it]);
                    term12[ir,it]=dfunc1[ir,it]
                     *cr[ir,it]-
                    func1[ir,it]*crr[ir,it]-
                    func1[ir,it]*(2/r[ir])*cr[ir,it];
                    term13[ir,it]=rm*m[ir,it]*c[ir,it]
                                   *(1-m[ir,it]);
                    term14[ir,it]=term11[ir,it]
                     +term12[ir,it]
                                      +term13[ir,it];
                    term21[ir,it]=(crr[ir,it]+(2/r[ir])
                                   *cr[ir,it]);
                    term22[ir,it]=func2[ir,it];
                    term23[ir,it]=-c[ir,it];
                    term24[ir,it]=term21[ir,it]
                     +term22[ir,it]
                                      +term23[ir,it];
                    term31[ir,it]=(1/tau)*(arr[ir,it]
                                   +(2/r[ir])*ar[ir,it])
                    term32[ir,it]=(1/tau)*func2[ir,it];
                    term33[ir,it]=-(1/tau)*a[ir,it];
                    term34[ir,it]=term31[ir,it]
                     +term32[ir,it]
                                      +term33[ir,it];}
    }
    }
```

The two final } conclude the fors in t and r.

- The computed terms `term11` to `temp34` are plotted against r and parametrically in t with the R utility `matplot`. For the first PDE, eqs. (1.3-1) and (1.3-4), the plotting is

```
#
# Plot PDE RHS, LHS terms
#
# term11
  par(mfrow=c(2,2));
  matplot(x=r,y=term11,type="l",xlab="r",ylab="term11(r,t)",
          xlim=c(rl,ru),lty=1,main="term11(r,t)",lwd=2,
          col="black",ylim=c(-5,5));
#
# term12
  matplot(x=r,y=term12,type="l",xlab="r",ylab="term12(r,t)",
          xlim=c(rl,ru),lty=1,main="term12(r,t)",lwd=2,
          col="black",ylim=c(-5,5));
#
# term13
  matplot(x=r,y=term13,type="l",xlab="r",ylab="term13(r,t)",
          xlim=c(rl,ru),lty=1,main="term13(r,t)",lwd=2,
          col="black",ylim=c(-5,5));
#
# term14
  matplot(x=r,y=term14,type="l",xlab="r",ylab="term14(r,t)",
          xlim=c(rl,ru),lty=1,main="term14(r,t)",lwd=2,
          col="black",ylim=c(-5,5));
```

A 2×2 matrix of plots is specified with `par(mfrow=c(2,2))`. The four plots have a common horizontal scale with `xlim=c(rl,ru)` and a common vertical scale with `ylim=c(-5,5)` so that the magnitudes of the terms can be compared by inspection. In other words, this scaling of the $x - y$ axes indicates (1) the relative contributions of the three PDE RHS terms and (2) the sum of the terms that equals the PDE LHS (derivative in t).

- For the second PDE, eqs. (1.3-2) and (1.3-5), the plotting is

```
#
# term21
  par(mfrow=c(2,2));
  matplot(x=r,y=term21,type="l",xlab="r",ylab="term21(r,t)",
          xlim=c(rl,ru),lty=1,main="term21(r,t)",lwd=2,
          col="black",ylim=c(-1,1));
#
# term22
  matplot(x=r,y=term22,type="l",xlab="r",ylab="term22(r,t)",
          xlim=c(rl,ru),lty=1,main="term22(r,t)",lwd=2,
```

```
                  col="black",ylim=c(-1,1));
    #
    # term23
      matplot(x=r,y=term23,type="l",xlab="r",ylab="term23(r,t)",
              xlim=c(rl,ru),lty=1,main="term23(r,t)",lwd=2,
              col="black",ylim=c(-1,1));
    #
    # term24
      matplot(x=r,y=term24,type="l",xlab="r",ylab="term24(r,t)",
              xlim=c(rl,ru),lty=1,main="term24(r,t)",lwd=2,
              col="black",ylim=c(-1,1));
```

The common vertical scaling in this case is ylim=c(-1,1).
- For the third PDE, eqs. (1.3-3) and (1.3-6), the plotting is

```
    #
    # term31
      par(mfrow=c(2,2));
      matplot(x=r,y=term31,type="l",xlab="r",ylab="term31(r,t)",
              xlim=c(rl,ru),lty=1,main="term31(r,t)",lwd=2,
              col="black",ylim=c(-1,1));
    #
    # term32
      matplot(x=r,y=term32,type="l",xlab="r",ylab="term32(r,t)",
              xlim=c(rl,ru),lty=1,main="term32(r,t)",lwd=2,
              col="black",ylim=c(-1,1));
    #
    # term33
      matplot(x=r,y=term33,type="l",xlab="r",ylab="term33(r,t)",
              xlim=c(rl,ru),lty=1,main="term33(r,t)",lwd=2,
              col="black",ylim=c(-1,1));
    #
    # term34
      matplot(x=r,y=term34,type="l",xlab="r",ylab="term34(r,t)",
              xlim=c(rl,ru),lty=1,main="term34(r,t)",lwd=2,
              col="black",ylim=c(-1,1));
```

The common vertical scaling in this case is ylim=c(-1,1).

This completes the discussion of the additional programming in the main program (Listing 4.1) to compute and display the PDE RHS and LHS terms, term11 to term34.

4.2.2 ODE/MOL routine

pde1a called by lsodes from the main program of Listing 4.1 is again in Listing 2.2.

4.2.3   Numerical, graphical output

The graphical output programmed in Listing 4.1 is discussed next. The terms for the first PDE of eqs. (1.3-1) and (1.3-4) is in Figure 4.1-1.

Figure 4.1-1 indicates that $\dfrac{\partial m}{\partial t}$ of term14 is determined by the linear (Fick's first law) diffusion of term11, that is, term12, term13 are negligible. $\dfrac{\partial m}{\partial t}$ of term14 is the same as in Figure 3.1-1.

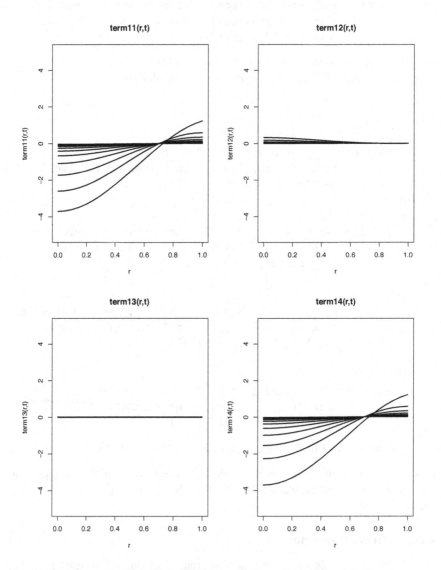

Figure 4.1-1 term11, term12, term13, term14, scaling ylim=c(-5,5).

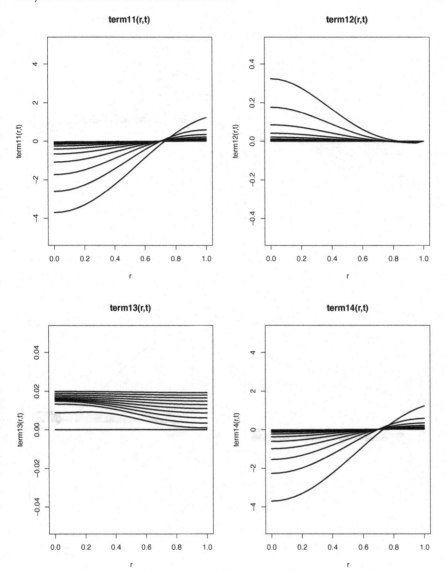

Figure 4.1-2 term12 scaled by ylim=c(-0.5,0.5) and term14 scaled by ylim=c(-0.05,0.05).

To confirm term12, term13 are negligible, Figure 4.1-2 has scaling ylim=c(-0.5,0.5) for term12 and ylim=c(-0.05,0.05) for term13.

Again, Figure 4.1-2 indicates that $\dfrac{\partial m}{\partial t}$ of term14 is determined by the linear (Fick's first law) diffusion of term11.

Figure 4.1-3 is a plot of term21, term22, term23, term24 with scaling ylim=c(-1,1).

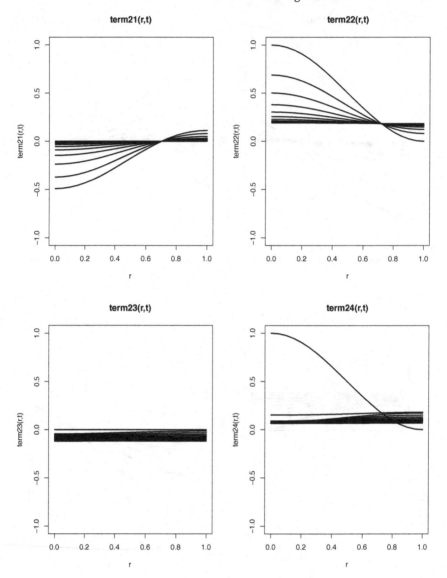

Figure 4.1-3 `term21, term22, term23 term24` with scaling `ylim=c(-1,1)`.

Figure 4.1-4 is a plot of `term31, term32, term33, term34` with scaling `ylim=c(-1,1)`.

Figures 4.1-3 and 4.1-4 are the same for the parameters in Listing 4.1, specifically, `tau=1`.

This completes the discussion of the calculation and display of the PDE RHS and LHS terms.

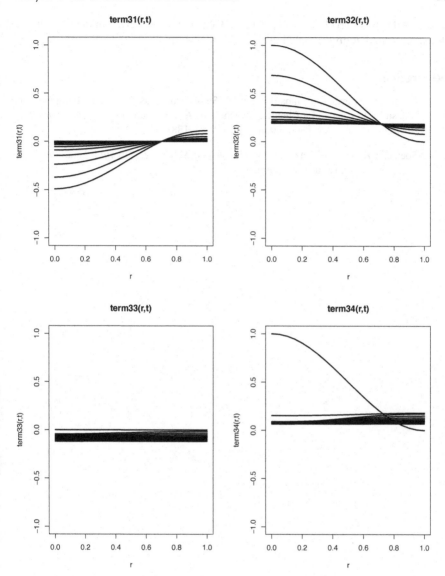

Figure 4.1-4   term31, term32, term33 term34 with scaling ylim=c(-1,1).

Summary and conclusion

The calculation and graphical display of the RHS and LHS terms of eqs. (1.3-1–1.3-6) provides an indication of the relative magnitudes of these terms. For the parameters of Listing 4.1, the derivatives $\dfrac{\partial m}{\partial t}, \dfrac{\partial c}{\partial t}, \dfrac{\partial a}{\partial t}$ are determined by the linear diffusion terms `term11`, `term21`, `term31`. Consideration is given in the next chapter to parameters that give nonnegligible chemotaxis terms.

References

1. Giunta, M., Carmela Lombardo, and M. Sammartino (2021), Pattern Formation and Transition to Chaos in a Chemotaxis Model of Acute Inflammation. *SIAM Journal on Applied Dynamical Systems*, **20**, no. 4, pp. 1844–1881.

2. Soetaert, K., J. Cash, and F. Mazzia (2012), *Solving Differential Equations in R*, Springer-Verlag, Heidelberg, Germany.

5 Sensitivity Analysis of the Chemotaxis PDE Model Parameters

5.1 Introduction

In Chapter 4, a methodology is explained and used for studying the RHS and LHS terms of eqs. (1.3-1–1.3-6) [1,2]. For the parameters of Listing 4.1, the linear (Fick's first law) diffusion terms of eqs. (1.3-1–1.3-6) essentially determined the LHS time derivative terms, and therefore the solutions $m(r,t), c(r,t), a(r,t)$. This conclusion suggests an increase in the chemotaxis coefficient χ in eqs. (1.3-1) and (1.3-4) to give a significant contribution to chemotaxis to $m(r,t), c(r,t), a(r,t)$, which is the first case considered next.

5.2 R routines for Case 1

The following change in χ of eqs. (1.3-1) and (1.3-4) indicates the sensitivity of the eqs. (1.3-1–1.3-6) solutions, $m(r,t), c(r,t), a(r,t)$, to the chemotaxis.

The change in Listing 4.1 is $\chi = 1$ to $\chi = 10$. The ODE/MOL routine is in Listing 2.2. The graphical output is in Figures 5.1-1–5.1-6.

Figure 5.1-1 can be compared with Figure 2.2-1a for an indication of the effect of the change in χ (1 to 10). Specifically, the change in the chemotxis term `term12` is significant, but the change in `term11` is also significant and opposite in sign so that the change in `term14`, and therefore $m(r,t)$, is small.

Figure 5.1-2 can be compared with Figure 2.2-2 for an indication of the effect of the change in χ (1 to 10). Specifically, the effect of this change on $c(r,t)$ is small as might be expected since the change in `term14` and $m(r,t)$ is small.

Figure 5.1-3 can be compared with Figure 2.2-3 for an indication of the effect of the change in χ (1 to 10). Specifically, the effect of this change on $a(r,t)$ is small as might be expected since the change in `term14` and $m(r,t)$ is small.

The plots for $\frac{\partial m}{\partial t}, \frac{\partial c}{\partial t}, \frac{\partial a}{\partial t}$ are not included here since they are also `term14`, `term24`, `term34` in Figure 5.1-4–5.1-6 next.

The scaling with `ylim=c(-5,5)` does not cover the variation in `term11`, `term12` (compare with Figure 4.1-1).

`term21`, `term22`, `term23`, `term24` in Figures 4.1-3 and 5.1-5 are similar.

`term31`, `term32`, `term33`, `term34` in Figures 4.1-4 and 5.1-6 are similar.

In summary, the change in χ (from 1 to 10) does not result in a substantial effect on $m(r,t), c(r,t), a(r,t)$.

DOI: 10.1201/9781003311201-5

The preceding case with $\chi = 10$ demonstrates that χ is a sensitive parameter in determining the chemotaxis `term12`. In fact, a change to $\chi = 25$ leads to a failure by `lsodes` of the t integration (solution) of the $3(21) = 63$ MOL ODEs. In other words, this is an example of a parameter value that leads to an unstable MOL calculation, and demonstrates the need for the careful selection of parameter values when developing a MOL formulation and computer implementation.

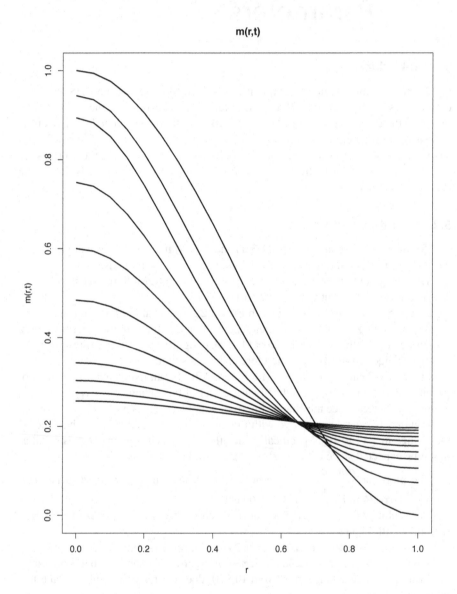

Figure 5.1-1 $m(r,t)$, $\chi = 10$.

c(r,t)

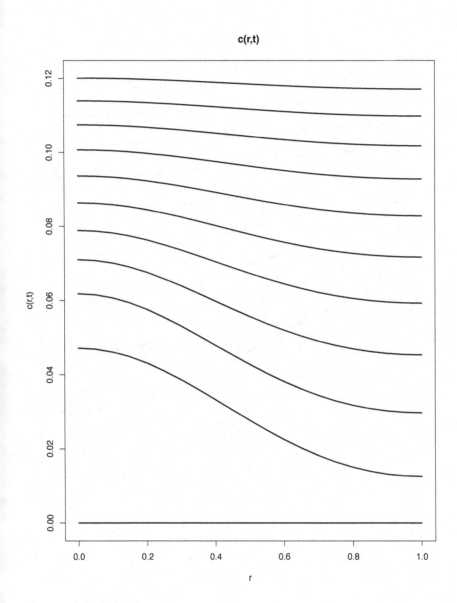

Figure 5.1-2 $c(r,t)$, $\chi = 10$.

a(r,t)

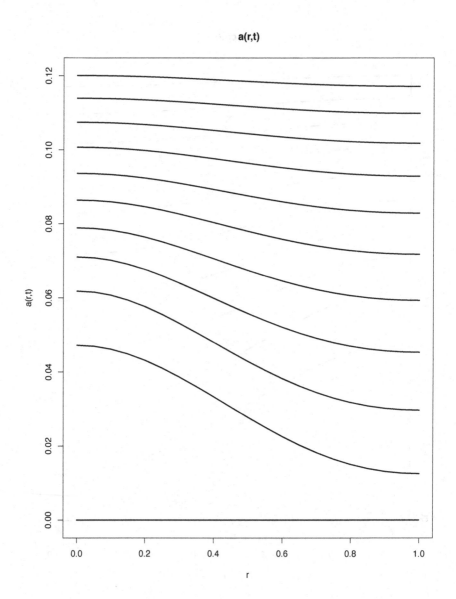

Figure 5.1-3 $a(r,t)$, $\chi = 10$.

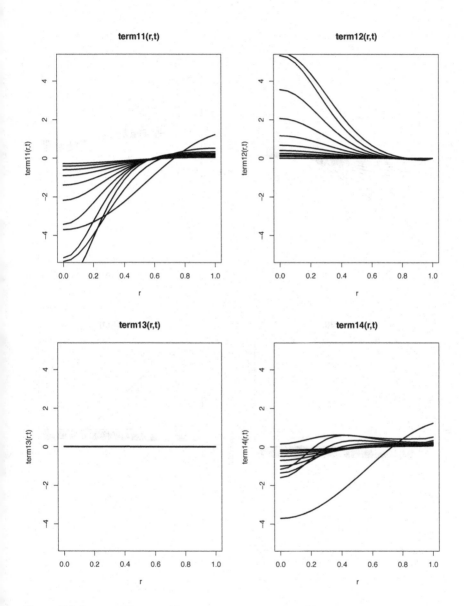

Figure 5.1-4 term11, term12, term13, term14, $\chi = 10$.

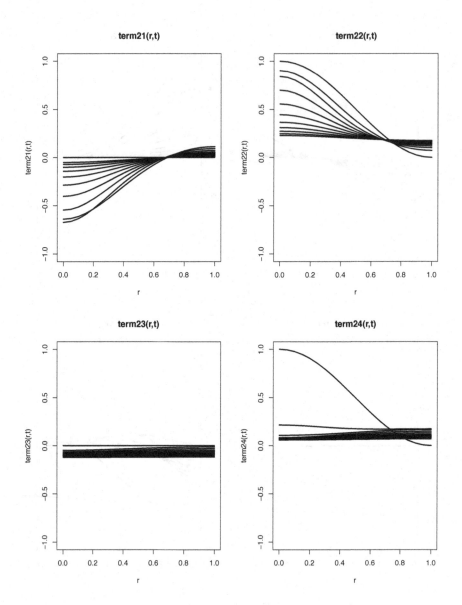

Figure 5.1-5 term21, term22, term23, term24, $\chi = 10$.

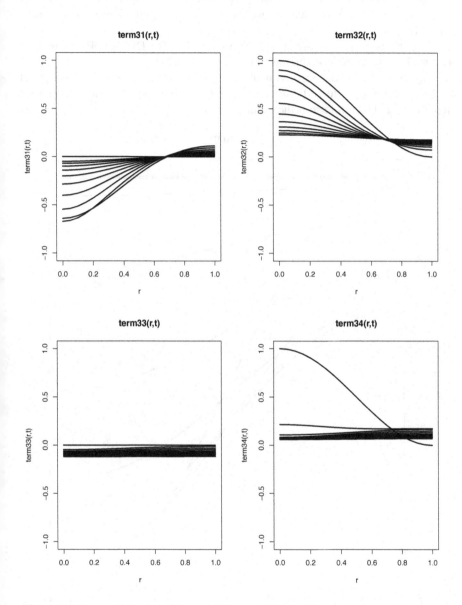

Figure 5.1-6 `term31, term32, term33, term34,` $\chi = 10$.

5.3   R routines for Case 2

The following change in r_m of eqs. (1.3-1) and (1.3-4) indicates the sensitivity of the eqs. (1.3-1–1.3-6) solutions, $m(r,t), c(r,t), a(r,t)$, to the logistic source term $r_m mc(1-m)$.

The change in Listing 4.1 is $r_m = 1$ to $r_m = 100$. The ODE/MOL routine is in Listing 2.2. The graphical output is in Figure 5.2-1–5.2-6.

Figure 5.2-1 can be compared with Figure 2.2-1a for an indication of the effect of the change in r_m (1 to 100). Specifically, the change in the source term `term13` is significant, and therefore so is the change in $m(r,t)$.

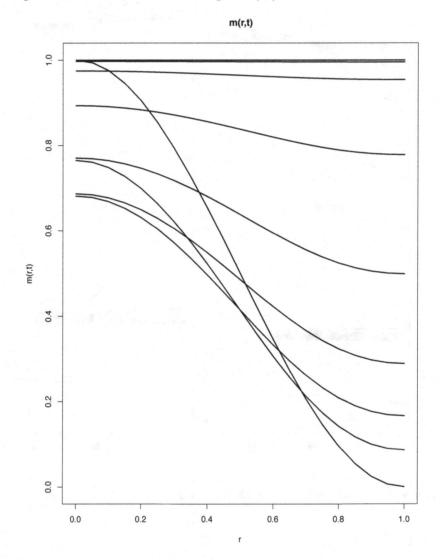

Figure 5.2-1 $m(r,t), r_m = 100$.

c(r,t)

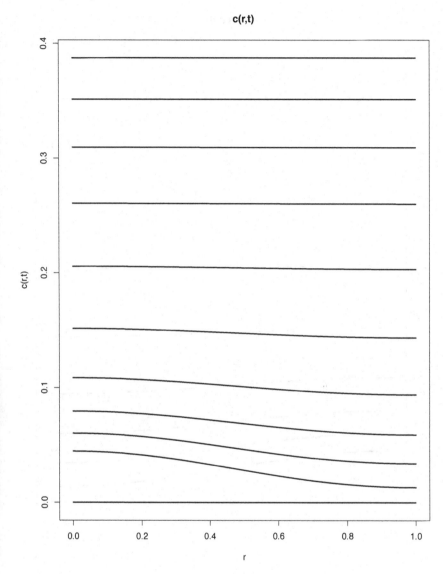

Figure 5.2-2 $c(r,t)$, $r_m = 100$.

Figure 5.2-2 can be compared with Figure 2.2-2 for an indication of the effect of the change in r_m (1 to 100). Specifically, the effect of this change is significant as a result of the coupling between eqs. (1.3-1) and (1.3-4) (for $m(r,t)$) and eqs. (1.3-2) and (1.3-5) (for $c(r,t)$).

Figure 5.2-3 can be compared with Figure 2.2-3 for an indication of the effect of the change in r_m (1 to 100). Specifically, the effect of this change is significant as a

a(r,t)

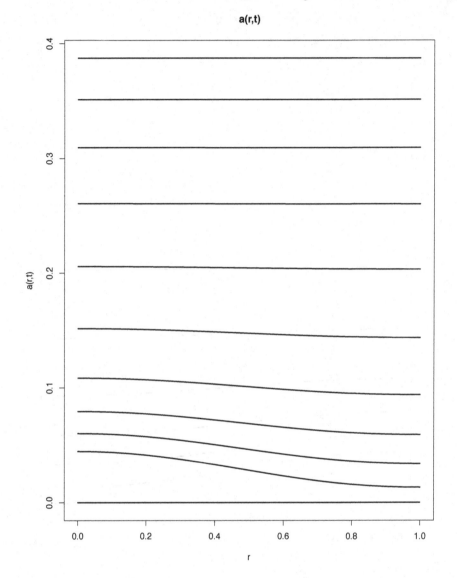

Figure 5.2-3 $a(r,t)$, $r_m = 100$.

result of the coupling between eqs. (1.3-1) and (1.3-4) (for $m(r,t)$) and eqs. (1.3-3) and (1.3-6) (for $a(r,t)$).

term11, term12, term13, term14 in Figures 4.1-1 and 5.2-4 are substantially different. The scaling with ylim=c(-5,5) covers the variation in term11 to term14.

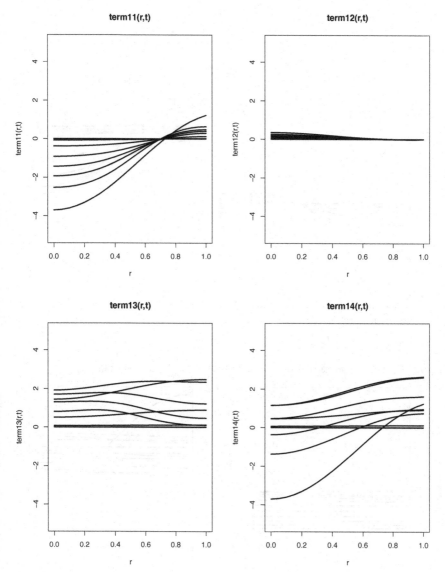

Figure 5.2-4 `term11, term12, term13, term14,` $r_m = 100$.

`term21, term22, term23, term24` in Figures 4.1-3 and 5.2-5 are substantially different. The scaling with `ylim=c(-1,1)` covers the variation in `term21` to `term24`.

`term31, term32, term33, term34` in Figures 4.1-4 and 5.2-6 are substantially different. The scaling with `ylim=c(-1,1)` covers the variation in `term31` to `term34`.

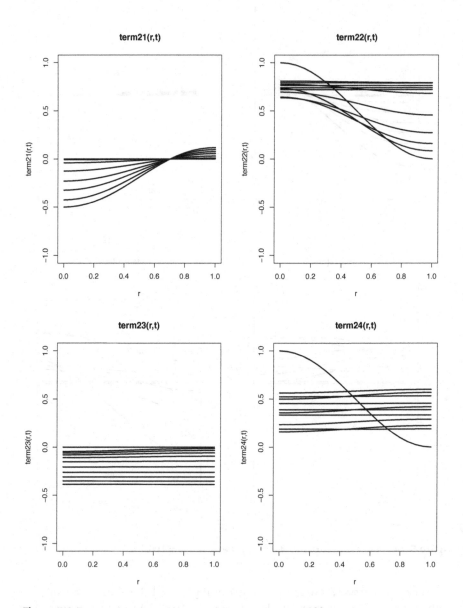

Figure 5.2-5 `term21, term22, term23, term24,` $r_m = 100$.

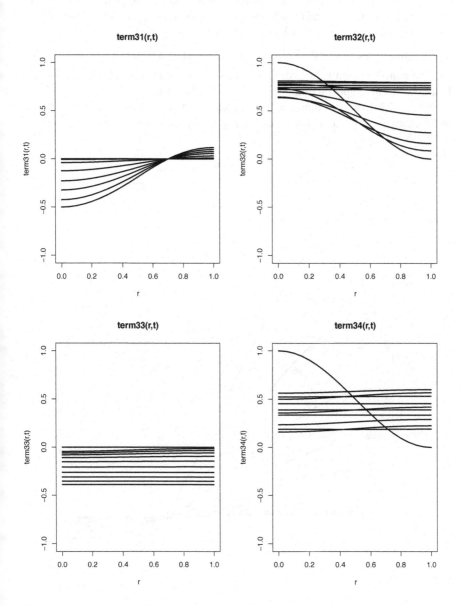

Figure 5.2-6 term31, term32, term33, term34, $r_m = 100$.

5.4   R routines for Case 3

As a concluding case of a parameter sensitivity analysis, a change in τ is considered. To reiterate, eqs. (1.3-2) and (1.3-5) for $c(r,t)$ are the same as eqs. (1.3-3) and (1.3-6) for $a(r,t)$ except for τ which is a time scale factor for $a(r,t)$. For example, for $\tau = 5$, $a(r,t)$ is five times slower than $c(r,t)$ ($\frac{\partial a}{\partial t}$ is reduced by a factor of $\frac{1}{5}$ relative to $\frac{\partial c}{\partial t}$).

The main program is in Listing 4.1 with $\tau = 5$ and the ODE/MOL routine is in Listing 2.2. The graphical output is in Figure 5.3-1–5.3-6.

Figure 5.2-1 can be compared with Figure 2.2-1a for an indication of the effect of the change in τ (1 to 5). The change is negligibly small since τ only affects $m(r,t)$ indirectly through $a(r,t)$.

m(r,t)

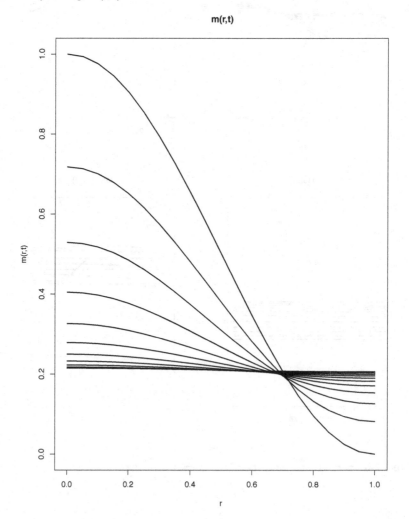

Figure 5.3-1 $m(r,t)$, $\tau = 5$.

c(r,t)

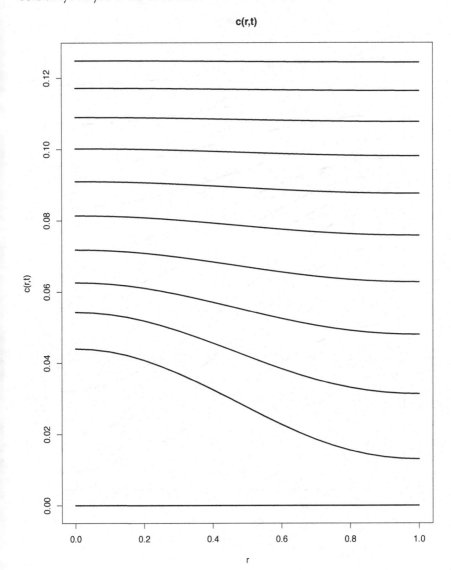

Figure 5.3-2 $c(r,t)$, $\tau = 5$.

Figure 5.3-2 can be compared with Figure 2.2-2 for an indication of the effect of the change in τ (1 to 5). The change is small since τ only affects $c(r,t)$ indirectly through $a(r,t)$.

Figure 5.2-3 can be compared with Figure 2.2-3 for an indication of the effect of the change in τ (1 to 5). $a(r,t)$ is significantly lower for $\tau = 5$ since $a(r,t)$ does not respond as rapidly in t.

a(r,t)

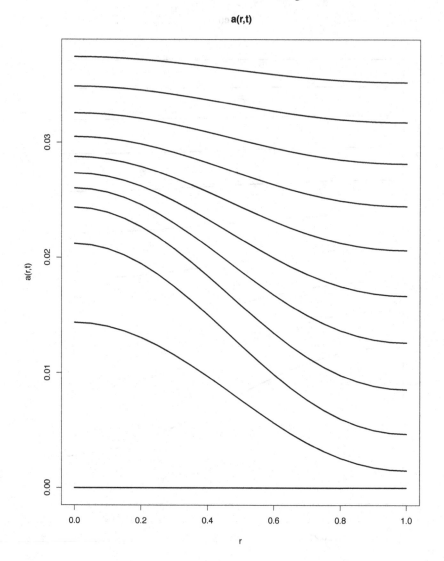

Figure 5.3-3 $a(r,t)$, $\tau = 5$.

term11, term12, term13, term14 in Figures 4.1-1 and 5.3-4 are essentially the same since τ affects mainly $a(r,t)$.

term21, term22, term23, term24 in Figures 4.1-3 and 5.3-5 are essentially the same since τ affects mainly $a(r,t)$.

term31, term32, term33, term34 in Figures 4.1-4 and 5.3-6 are substantially different. $a(r,t)$ is significantly lower for $\tau = 5$ since it does not respond as rapidly (depart from the zero IC) in t.

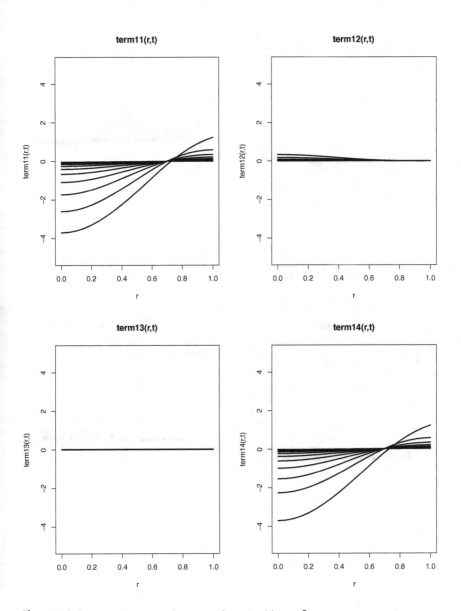

Figure 5.3-4 `term11, term12, term13, term14,` $\tau = 5$.

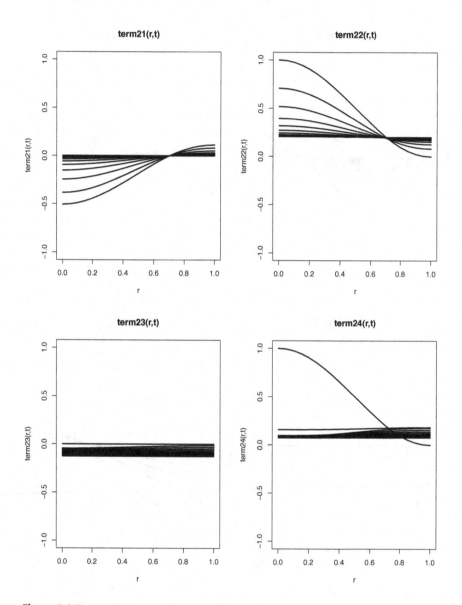

Figure 5.3-5 term21, term22, term23, term24, $\tau = 5$.

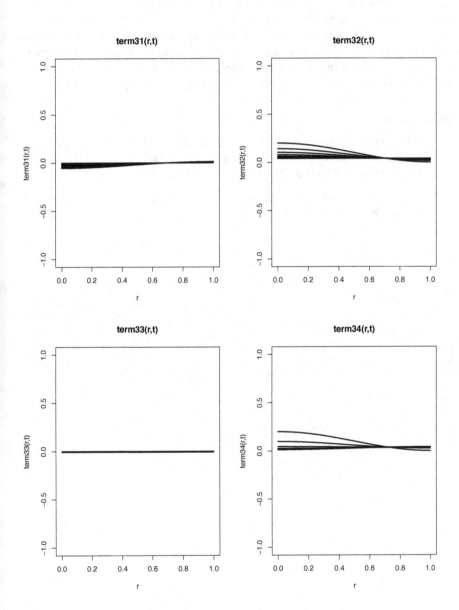

Figure 5.3-6 `term31, term32, term33, term34,` $\tau = 5$.

Summary and conclusion

The three cases (for χ, r_m, τ) demonstrate a methodology for studying parameter sensitivity based principally on the graphical output of the PDE RHS and LHS terms. This type of analysis clearly explains the source of the PDE solution features $(m(r,t), c(r,t), a(r,t)$ in this case) and is essential in the development of a ODE/PDE-based computer-based model. For the model of eqs. (1.3-1–1.3-6), an explanation of the chemotaxis generation of autoimmune inflammation is the principal result.

References

1. Giunta, M., Carmela Lombardo, and M. Sammartino (2021), Pattern Formation and Transition to Chaos in a Chemotaxis Model of Acute Inflammation. *SIAM Journal on Applied Dynamical Systems*, **20**, no. 4, pp. 1844–1881.

2. Soetaert, K., J. Cash, and F. Mazzia (2012), *Solving Differential Equations in R*, Springer-Verlag, Heidelberg, Germany.

6 Extensions of the Three PDE Chemotaxis Model

6.1 Introduction

The PDE model of eqs. (1.3-1–1.3-6) is now extended to include the effect of a postulated anti-inflammatory therapeutic drug.

6.2 PDE chemotaxis model with anti-inflammatory drug

Eqs. (1.3-1–1.3-6) and (1.4-1–1.4-3) are restated with the addition of terms to represent the effect of a therapeutic anti-inflammatory drug [1, 2].

$$\frac{\partial m}{\partial t} = D\left(\frac{\partial^2 m}{\partial r^2} + \frac{2}{r}\frac{\partial m}{\partial r}\right)$$

$$-\frac{\partial \chi \frac{m}{(1+\alpha c)^2}}{\partial r}\frac{\partial c}{\partial r} - \chi\frac{m}{(1+\alpha c)^2}\frac{\partial^2 c}{\partial r^2} - \chi\frac{m}{(1+\alpha c)^2}\frac{2}{r}\frac{\partial c}{\partial r}$$

$$+ r_m mc(1-m) \tag{6.1-1}$$

$$\frac{\partial c}{\partial t} = \frac{\partial^2 c}{\partial r^2} + \frac{2}{r}\frac{\partial c}{\partial r} + \frac{m}{1+\beta a^p} - c - k_c c \tag{6.1-2}$$

$$\frac{\partial a}{\partial t} = \frac{1}{\tau}\left(\frac{\partial^2 a}{\partial r^2} + \frac{2}{r}\frac{\partial a}{\partial r} + \frac{m}{1+\beta a^p} - a - k_a a\right) \tag{6.1-3}$$

Eqs. (6.1-1–6.1-6) for $\dfrac{\partial m(r=0,t)}{\partial t}$, $\dfrac{\partial c(r=0,t)}{\partial t}$, $\dfrac{\partial a(r=0,t)}{\partial t}$ are (with BCs (1.4-1–1.4-3))

$$\frac{\partial m}{\partial t} = 3D\frac{\partial^2 m}{\partial r^2} - 3\chi\frac{m}{(1+\alpha c)^2}\frac{\partial^2 c}{\partial r^2} + r_m mc(1-m) \tag{6.1-4}$$

$$\frac{\partial c}{\partial t} = 3\frac{\partial^2 c}{\partial r^2} + \frac{m}{1+\beta a^p} - c - k_c c \tag{6.1-5}$$

$$\frac{\partial a}{\partial t} = \frac{1}{\tau}\left(3\frac{\partial^2 a}{\partial r^2} + \frac{m}{1+\beta a^p} - a - k_a a\right) \tag{6.1-6}$$

In particular, eqs. (6.1-2, 6.1-5) have the additional term $-k_c c$ for the effect of the anti-inflammatory drug on the cytokine concentration $c(r,t)$ and eqs. (6.1-3, 6.1-6)

DOI: 10.1201/9781003311201-6

have the additional term $-k_a a$ for the effect of the anti-inflammatory drug on the cytokine concentration $a(r,t)$. The BCs for eqs. (6.1-1–6.1-6) remain as in eqs. (1.4-1–1.4-3), that is, homogeneous (zero, no flux) Neumann BCs.

$$\frac{\partial m(r = r_l = 0, t)}{\partial r} = \frac{\partial m(r = r_u = 1, t)}{\partial r} = 0 \qquad (6.2\text{-}1)$$

$$\frac{\partial c(r = r_l = 0, t)}{\partial r} = \frac{\partial c(r = r_u = 1, t)}{\partial r} = 0 \qquad (6.2\text{-}2)$$

$$\frac{\partial a(r = r_l = 0, t)}{\partial r} = \frac{\partial a(r = r_u = 1, t)}{\partial r} = 0 \qquad (6.2\text{-}3)$$

The R routines for eqs. (6.1-1–6.1-6) and (6.2-1–6.2-3) are straightforward extensions of Listings 2.1 and 2.2.

6.2.1   Main program

Only the changes in Listing 2.1 are indicated here.

- The setwd is changed to designate Chapter 6.

```
#
# Access functions for numerical solution
  setwd("f:/inflammation/chap6");
  source("pde1a.R");
  source("dss004.R");
  source("dss044.R");
```

- The rate constants k_c, k_a are added to the parameters.

```
#
# Parameters
  D=0.25;
  rm=1;
  chi=1;
  alpha=1;
  beta=1;
  p=1;
  tau=1;
  m0=1;
  kc=1;
  ka=1;
```

- The IC $m(r, t = 0)$ is a half cos wave.

```
#
# Initial conditions
  u0=rep(0,3*nr);
  for(ir in 1:nr){
    u0[ir]      =0.5*(1+cos(pi*(r[ir]-rl)/(ru-rl)));
    u0[ir+nr]   =0;
    u0[ir+2*nr] =0;
  }
```

- The plotting is simplified.

```
#
# Plot PDE solutions
#
# m
  par(mfrow=c(1,1));
  matplot(x=r,y=m,type="l",xlab="r",ylab="m(r,t)",
          xlim=c(rl,ru),lty=1,main="m(r,t)",lwd=2,
          col="black");
  persp(r,tout,m,theta=60,phi=45,
        xlim=c(rl,ru),ylim=c(t0,tf),zlim=c(0,1.1),
        xlab="r",ylab="t",zlab="m(r,t)");
#
# c
  par(mfrow=c(1,1));
  matplot(x=r,y=c,type="l",xlab="r",ylab="c(r,t)",
          xlim=c(rl,ru),lty=1,main="c(r,t)",lwd=2,
          col="black");
#
# a
  par(mfrow=c(1,1));
  matplot(x=r,y=a,type="l",xlab="r",ylab="a(r,t)",
          xlim=c(rl,ru),lty=1,main="a(r,t)",lwd=2,
          col="black");
```

Listing 6.1 Main program revisions for eqs. (6.1-1–6.1-6)

6.2.2 MOL/ODE routine

Only the changes in Listing 2.2 are indicated here. Specifically, the MOL programming for $\frac{\partial c}{\partial t}$=ct[ir], $\frac{\partial a}{\partial t}$=at[ir] includes the rate terms $-k_c c, -k_a a$.

```
#
# PDEs
  mt=rep(0,nr);ct=rep(0,nr);at=rep(0,nr);
  for(ir in 1:nr){
    if(ir==1){
      mt[ir]=3*D*mrr[ir]-3*func1[ir]*crr[ir]+
             rm*m[ir]*c[ir]*(1-m[ir]);
      ct[ir]=3*crr[ir]+func2[ir]-c[ir]-kc*c[ir];
      at[ir]=(1/tau)*(3*arr[ir]+func2[ir]-a[ir]-ka*a[ir]);
    }
    if(ir==nr){
      mt[ir]=D*mrr[ir]-func1[ir]*crr[ir]+
             rm*m[ir]*c[ir]*(1-m[ir]);
      ct[ir]=crr[ir]+func2[ir]-c[ir]-kc*c[ir];
      at[ir]=(1/tau)*(arr[ir]+func2[ir]-a[ir]-ka*a[ir]);
    }
    if((ir>1)&(ir<nr)){
      mt[ir]=D*(mrr[ir]+(2/r[ir])*mr[ir])-
             dfunc1[ir]*cr[ir]-func1[ir]*crr[ir]-
             func1[ir]*(2/r[ir])*cr[ir]+
             rm*m[ir]*c[ir]*(1-m[ir]);
      ct[ir]=(crr[ir]+(2/r[ir])*cr[ir])+func2[ir]-
             c[ir]-kc*c[ir];
      at[ir]=(1/tau)*(arr[ir]+(2/r[ir])*ar[ir]+
             func2[ir]-a[ir]-ka*a[ir]);
    }
  }
```

Listing 6.2 MOL/ODE routine revisions for eqs. (6.1-1–6.1-6)

Abbreviated output from Listings 6.1 and 6.2 is shown in Table 6.1.

6.2.3 Numerical, graphical output

The effect of the terms $-k_c c, -k_a a$ is to reduce the concentrations of the cytokines. For example, a comparison of the $t = 1$ output from Tables 2.2 and 6.1.

```
Table 2.2
```

t	r	m(r,t)	c(r,t)	a(r,t)
1.00	0.00	2.143e-01	1.182e-01	1.182e-01
1.00	0.05	2.142e-01	1.182e-01	1.182e-01
1.00	0.10	2.141e-01	1.181e-01	1.181e-01
.		.		.
.		.		.
.		.		.

Table 6.1
Numerical Output for eqs. (6.1-1–6.1-6)

```
[1] 11

[1] 64

   t      r      m(r,t)        c(r,t)        a(r,t)
 0.00   0.00   1.000e+00    0.000e+00    0.000e+00
 0.00   0.05   9.938e-01    0.000e+00    0.000e+00
 0.00   0.10   9.755e-01    0.000e+00    0.000e+00
          .              .
          .              .
          .              .

Output for t=0.15 to 0.85 is removed
          .              .
          .              .
          .              .

 0.00   0.90   2.447e-02    0.000e+00    0.000e+00
 0.00   0.95   6.156e-03    0.000e+00    0.000e+00
 0.00   1.00   0.000e+00    0.000e+00    0.000e+00

   t      r      m(r,t)        c(r,t)        a(r,t)
 1.00   0.00   2.117e-01    8.186e-02    8.186e-02
 1.00   0.05   2.116e-01    8.185e-02    8.185e-02
 1.00   0.10   2.115e-01    8.185e-02    8.185e-02
          .              .
          .              .
          .              .

Output for t=0.15 to 0.85 is removed
          .              .
          .              .
          .              .

 1.00   0.90   2.033e-01    8.141e-02    8.141e-02
 1.00   0.95   2.032e-01    8.141e-02    8.141e-02
 1.00   1.00   2.031e-01    8.141e-02    8.141e-02

ncall =    172
```

```
Output for t=0.15 to 0.85 is removed
              .                    .
              .                    .
              .                    .

1.00  0.90    2.058e-01    1.177e-01    1.177e-01
1.00  0.95    2.057e-01    1.177e-01    1.177e-01
1.00  1.00    2.056e-01    1.177e-01    1.177e-01
```

Table 6.1

t	r	m(r,t)	c(r,t)	a(r,t)
1.00	0.00	2.117e-01	8.186e-02	8.186e-02
1.00	0.05	2.116e-01	8.185e-02	8.185e-02
1.00	0.10	2.115e-01	8.185e-02	8.185e-02
	.	.	.	
	.	.	.	
	.	.	.	

```
Output for t=0.15 to 0.85 is removed
              .                    .
              .                    .
              .                    .

1.00  0.90    2.033e-01    8.141e-02    8.141e-02
1.00  0.95    2.032e-01    8.141e-02    8.141e-02
1.00  1.00    2.031e-01    8.141e-02    8.141e-02
```

This indicates little change in $m(r,t)$, but substantial reductions in $c(r,t), a(r,t)$ indicating the effectiveness of the postulated anti-inflammatory drug.

The conclusion also follows from comparison of Figures (2.2-1a, 2.2-1b to 2.2-3) and (6.1-1a, 6.1-1b to 6.1-3).

Figure 6.1-1a indicates a small change in $m(r,t)$ from Figure 2.2-1a.

Comparison of Figures 2.2-2 and 6.1-2 indicates that $c(r,t)$ is substantially lowered.

Comparison of Figures 2.2-3 and 6.1-3 indicates that $a(r,t)$ is substantially lowered.

In summary, the effectiveness of the postulated anti-inflammatory drug is demonstrated for the values $k_c = k_a = 1$ in eqs. (6.1-1–6.1-6).

m(r,t)

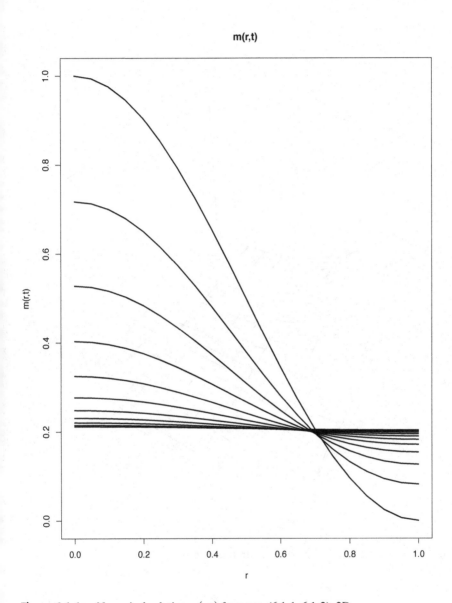

Figure 6.1-1a   Numerical solution $m(r,t)$ from eqs. (6.1-1, 6.1-2), 2D.

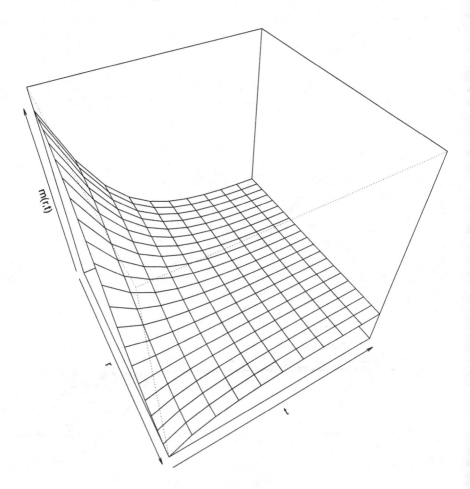

Figure 6.1-1b   Numerical solution $m(r,t)$ from eqs. (6.1-1, 6.1-2), 3D.

c(r,t)

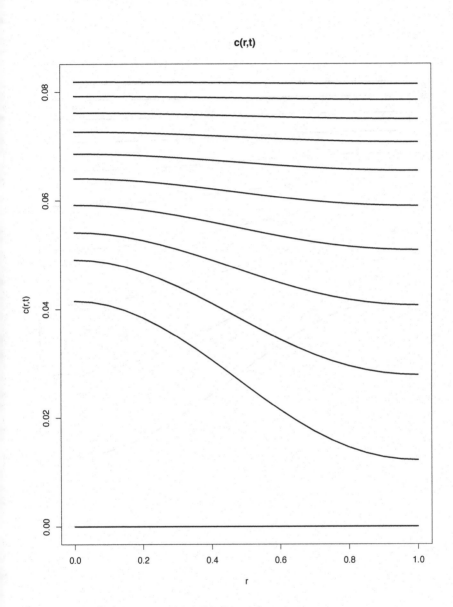

Figure 6.1-2 $c(r,t)$ from eqs. (6.1-1, 6.1-2).

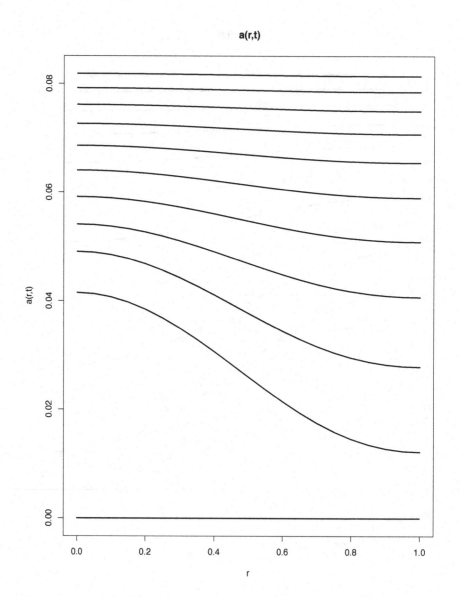

Figure 6.1-3 $a(r,t)$ from eqs. (6.1-1, 6.1-2).

6.3 Multicomponent extension of PDE chemotaxis model

The following example of the extension of the PDE chemotaxis inflammation starts from the basic three PDE model of Chapter 2. The dependent variables $m(r,t), c(r,t), (a,r)$ are defined by eqs. (1.3-1–1.3-6) and (1.4-1–1.4-3), and implemented in the main program and ODE/MOL routines of Listing 2.1. In the following development, a second set of PDEs is added to the R routines of Listings 2.1 and 2.2. Two models are then defined by the following six PDEs.

The first three PDEs are eqs. (1.3-1–1.3-6) with $m(r,t), c(r,t), a(r,t)$ designated as $m_1(r,t), c_1(r,t), a_1(r,t)$.

$$\frac{\partial m_1}{\partial t} = D_1 \left(\frac{\partial^2 m_1}{\partial r^2} + \frac{2}{r} \frac{\partial m_1}{\partial r} \right)$$

$$- \frac{\partial \chi_1 \frac{m_1}{(1+\alpha_1 c_1)^2} \partial c_1}{\partial r} \frac{\partial c_1}{\partial r} - \chi_1 \frac{m_1}{(1+\alpha c_1)^2} \frac{\partial^2 c_1}{\partial r^2} - \chi_1 \frac{m_1}{(1+\alpha_1 c_1)^2} \frac{2}{r} \frac{\partial c_1}{\partial r}$$

$$+ r_m m_1 c_1 (1 - m_1) \tag{6.3-1}$$

$$\frac{\partial c_1}{\partial t} = \frac{\partial^2 c_1}{\partial r^2} + \frac{2}{r} \frac{\partial c_1}{\partial r} + \frac{m_1}{1 + \beta_1 a_1^{p_1}} - c_1 \tag{6.3-2}$$

$$\frac{\partial a_1}{\partial t} = \frac{1}{\tau_1} \left(\frac{\partial^2 a_1}{\partial r^2} + \frac{2}{r} \frac{\partial a_1}{\partial r} + \frac{m_1}{1 + \beta_1 a_1^{p_1}} - a \right) \tag{6.3-3}$$

Eqs. (6.3-1–6.3-6) for $\frac{\partial m_1(r=0,t)}{\partial t}, \frac{\partial c_1(r=0,t)}{\partial t}, \frac{\partial a_1(r=0,t)}{\partial t}$ are (with BCs (6.4))

$$\frac{\partial m_1}{\partial t} = 3 D_1 \frac{\partial^2 m_1}{\partial r^2} - 3 \chi_1 \frac{m_1}{(1+\alpha c_1)^2} \frac{\partial^2 c_1}{\partial r^2} + r_m m_1 c_1 (1 - m_1) \tag{6.3-4}$$

$$\frac{\partial c_1}{\partial t} = 3 \frac{\partial^2 c_1}{\partial r^2} + \frac{m_1}{1 + \beta_1 a_1^{p_1}} - c_1 \tag{6.3-5}$$

$$\frac{\partial a_1}{\partial t} = \frac{1}{\tau_1} \left(3 \frac{\partial^2 a_1}{\partial r^2} + \frac{m_1}{1 + \beta_1 a_1^{p_1}} - a_1 \right) \tag{6.3-6}$$

The BCs for eqs. (6.3-1–6.3-6) are

$$\frac{\partial m_1(r=r_l=0,t)}{\partial r} = \frac{\partial m_1(r=r_u=1,t)}{\partial r} = 0 \tag{6.4-1}$$

$$\frac{\partial c_1(r=r_l=0,t)}{\partial r} = \frac{\partial c_1(r=r_u=1,t)}{\partial r} = 0 \tag{6.4-2}$$

$$\frac{\partial a_1(r=r_l=0,t)}{\partial r} = \frac{\partial a_1(r=r_u=1,t)}{\partial r} = 0 \tag{6.4-3}$$

The second set of three PDEs are eqs. (1.3-1–1.3-6) with $m(r,t), c(r,t), a(r,t)$ designated as $m_2(r,t)$, $c_2(r,t)$, $a_2(r,t)$.

$$\frac{\partial m_2}{\partial t} = D_2 \left(\frac{\partial^2 m_2}{\partial r^2} + \frac{2}{r}\frac{\partial m_2}{\partial r} \right)$$

$$- \frac{\partial \chi_2 \frac{m_2}{(1+\alpha_2 c_2)^2}}{\partial r} \frac{\partial c_2}{\partial r} - \chi_2 \frac{m_2}{(1+\alpha c_2)^2}\frac{\partial^2 c_2}{\partial r^2} - \chi_2 \frac{m_2}{(1+\alpha_2 c_2)^2}\frac{2}{r}\frac{\partial c_2}{\partial r}$$

$$+ r_m m_2 c_2(1-m_2) \tag{6.5-1}$$

$$\frac{\partial c_2}{\partial t} = \frac{\partial^2 c_2}{\partial r^2} + \frac{2}{r}\frac{\partial c_2}{\partial r} + \frac{m_2}{1+\beta_2 a_2^{p_2}} - c_2 \tag{6.5-2}$$

$$\frac{\partial a_2}{\partial t} = \frac{1}{\tau_2}\left(\frac{\partial^2 a_2}{\partial r^2} + \frac{2}{r}\frac{\partial a_2}{\partial r} + \frac{m_2}{1+\beta_2 a_2^{p_2}} - a_2 \right) \tag{6.5-3}$$

Eqs. (6.5-1–6.5-6) for $\dfrac{\partial m_2(r=0,t)}{\partial t}$, $\dfrac{\partial c_2(r=0,t)}{\partial t}$, $\dfrac{\partial a_2(r=0,t)}{\partial t}$ are (with BCs (6.6))

$$\frac{\partial m_2}{\partial t} = 3D_2 \frac{\partial^2 m_2}{\partial r^2} - 3\chi_2 \frac{m_2}{(1+\alpha c_2)^2}\frac{\partial^2 c_2}{\partial r^2} + r_m m_2 c_2(1-m_2) \tag{6.5-4}$$

$$\frac{\partial c_2}{\partial t} = 3\frac{\partial^2 c_2}{\partial r^2} + \frac{m_2}{1+\beta_2 a_2^{p_2}} - c_2 \tag{6.5-5}$$

$$\frac{\partial a_2}{\partial t} = \frac{1}{\tau_2}\left(3\frac{\partial^2 a_2}{\partial r^2} + \frac{m_2}{1+\beta_2 a_2^{p_2}} - a_2 \right) \tag{6.5-6}$$

The BCs for eqs. (6.5-1–6.5-6) are

$$\frac{\partial m_2(r=r_l=0,t)}{\partial r} = \frac{\partial m_2(r=r_u=1,t)}{\partial r} = 0 \tag{6.6-1}$$

$$\frac{\partial c_2(r=r_l=0,t)}{\partial r} = \frac{\partial c_2(r=r_u=1,t)}{\partial r} = 0 \tag{6.6-2}$$

$$\frac{\partial a_2(r=r_l=0,t)}{\partial r} = \frac{\partial a_2(r=r_u=1,t)}{\partial r} = 0 \tag{6.6-3}$$

6.3.1   Main program

The two models of eqs. (6.1-1–6.6-3) are programmed in the following routines, starting with a main program. The intention of this case study is to demonstrate a methodology for implementing two models to execute in parallel in time t (simultaneously). This arrangement then facilitates a comparison of the models.

```
#
# Two three PDE chemotaxis inflammation models
#
```

```
# Delete previous workspaces
  rm(list=ls(all=TRUE))
#
# Access ODE integrator
  library("deSolve");
#
# Access functions for numerical solution
  setwd("f:/inflammation/chap6");
  source("pde1b.R");
  source("dss004.R");
  source("dss044.R");
#
# Number of components
  nc=1;
#
# Parameters
  D1=0.25;
  rm1=1;
  chi1=1;
  alpha1=1;
  beta1=1;
  p1=1;
  tau1=1;
  m10=1;
#
# Two components
  if(nc==2){
  D2=0.25;
  rm2=1;
  chi2=1;
  alpha2=1;
  beta2=1;
  p2=1;
  tau2=1;
  m20=1;
  }
#
# Grid (in r)
  nr=21;rl=0;ru=1
  r=seq(from=rl,to=ru,by=(ru-rl)/(nr-1));
#
# Independent variable for ODE integration
  t0=0;tf=1;nout=11;
  tout=seq(from=t0,to=tf,by=(tf-t0)/(nout-1));
#
```

```
# Initial conditions
  u0=rep(0,3*nr*nc);
  for(ir in 1:nr){
    u0[ir]    =0.5*(1+cos(pi*(r[ir]-rl)/(ru-rl)));
    u0[ir+nr] =0;
    u0[ir+2*nr]=0;
  }
  if(nc==2){
  for(ir in 1:nr){
    u0[ir+3*nr]=0.5*(1+cos(pi*(r[ir]-rl)/(ru-rl)));
    u0[ir+4*nr]=0;
    u0[ir+5*nr]=0;
  }
  }
  ncall=0;
#
# ODE integration
  out=lsodes(y=u0,times=tout,func=pde1b,
     sparsetype ="sparseint",rtol=1e-6,
     atol=1e-6,maxord=5);
  nrow(out)
  ncol(out)
#
# Arrays for plotting numerical solution
  m1=matrix(0,nrow=nr,ncol=nout);
  c1=matrix(0,nrow=nr,ncol=nout);
  a1=matrix(0,nrow=nr,ncol=nout);
  for(it in 1:nout){
    for(ir in 1:nr){
      m1[ir,it]=out[it,ir+1];
      c1[ir,it]=out[it,ir+1+nr];
      a1[ir,it]=out[it,ir+1+2*nr];
    }
  }
  if(nc==2){
  m2=matrix(0,nrow=nr,ncol=nout);
  c2=matrix(0,nrow=nr,ncol=nout);
  a2=matrix(0,nrow=nr,ncol=nout);
  for(it in 1:nout){
    for(ir in 1:nr){
      m2[ir,it]=out[it,ir+1+3*nr];
      c2[ir,it]=out[it,ir+1+4*nr];
      a2[ir,it]=out[it,ir+1+5*nr];
    }
  }
```

```
      }
  #
  # Display numerical solution
    for(it in 1:nout){
      if((it==1)|(it==nout)){
        cat(sprintf("\n t   r   m1(r,t)   c1(r,t) a1(r,t)\n"));
        for(ir in 1:nr){
          cat(sprintf("%6.2f%6.2f%12.3e%12.3e%12.3e\n",
            tout[it],r[ir],m1[ir,it],c1[ir,it],a1[ir,it]));
        }
      }
    }
    if(nc==2){
    for(it in 1:nout){
      if((it==1)|(it==nout)){
        cat(sprintf("\n t r  m2(r,t)   c2(r,t)   a2(r,t)\n"));
        for(ir in 1:nr){
          cat(sprintf("%6.2f%6.2f%12.3e%12.3e%12.3e\n",
            tout[it],r[ir],m2[ir,it],c2[ir,it],a2[ir,it]));
        }
      }
    }
    }
    }
  #
  # Calls to ODE routine
    cat(sprintf("\n\n ncall = %5d\n\n",ncall));
  #
  # Plot PDE solutions
  #
  # m1
    par(mfrow=c(1,1));
    matplot(x=r,y=m1,type="l",xlab="r",ylab="m1(r,t)",
            xlim=c(rl,ru),lty=1,main="m1(r,t)",lwd=2,
            col="black");
    persp(r,tout,m1,theta=60,phi=45,
          xlim=c(rl,ru),ylim=c(t0,tf),zlim=c(0,1.1),
          xlab="r",ylab="t",zlab="m1(r,t)");
  #
  # c1
    par(mfrow=c(1,1));
    matplot(x=r,y=c1,type="l",xlab="r",ylab="c1(r,t)",
            xlim=c(rl,ru),lty=1,main="c1(r,t)",lwd=2,
            col="black");
  #
  # a1
```

```
par(mfrow=c(1,1));
matplot(x=r,y=a1,type="l",xlab="r",ylab="a1(r,t)",
        xlim=c(rl,ru),lty=1,main="a1(r,t)",lwd=2,
        col="black");
if(nc==2){
#
# m2
par(mfrow=c(1,1));
matplot(x=r,y=m2,type="l",xlab="r",ylab="m2(r,t)",
        xlim=c(rl,ru),lty=1,main="m2(r,t)",lwd=2,
        col="black");
persp(r,tout,m2,theta=60,phi=45,
      xlim=c(rl,ru),ylim=c(t0,tf),zlim=c(0,1.1),
      xlab="r",ylab="t",zlab="m2(r,t)");
#
# c2
par(mfrow=c(1,1));
matplot(x=r,y=c2,type="l",xlab="r",ylab="c2(r,t)",
        xlim=c(rl,ru),lty=1,main="c2(r,t)",lwd=2,
        col="black");
#
# a2
par(mfrow=c(1,1));
matplot(x=r,y=a2,type="l",xlab="r",ylab="a2(r,t)",
        xlim=c(rl,ru),lty=1,main="a2(r,t)",lwd=2,
        col="black");
}
```

Listing 6.3 Main program for eqs. (6.1-1–6.6-3)

We can note the following details about Listing 6.3 (with some repetition of the discussion of Listing 2.1 so that this explanation is self-contained).

- Previous workspaces are deleted.

  ```
  #
  # Two three PDE chemotaxis inflammation models
  #
  # Delete previous workspaces
    rm(list=ls(all=TRUE))
  ```

- The R ODE integrator library deSolve is accessed [2]. Then the directory with the files for the solution of eqs. (6.1-1–6.6-3) is designated. Note that setwd (set working directory) uses / rather than the usual \.

  ```
  #
  # Access ODE integrator
    library("deSolve");
  ```

```
#
# Access functions for numerical solution
  setwd("f:/inflammation/chap6);
  source("pde1b.R");
  source("dss004.R");
  source("dss044.R");
```

dss004, dss044 are library routines for the calculation of spatial first and second derivatives. These routines are listed and explained in Appendix A1.

- The number of component models is defined (nc = 1,2).

```
#
# Number of components
  nc=1;
```

- The parameters for the two models are defined numerically.

```
#
# Parameters
  D1=0.25;
  rm1=1;
  chi1=1;
  alpha1=1;
  beta1=1;
  p1=1;
  tau1=1;
  m10=1;
#
# Two components
  if(nc==2){
  D2=0.25;
  rm2=1;
  chi2=1;
  alpha2=1;
  beta2=1;
  p2=1;
  tau2=1;
  m20=1;
  }
```

The branch for the second model is if(nc==2).

- A spatial grid for eqs. (6.3-1–6.6-3) is defined with 21 points so that $r = 0,0.05,\ldots,1$.

```
#
# Grid (in r)
  nr=21;rl=0;ru=1
  r=seq(from=rl,to=ru,by=(ru-rl)/(nr-1));
```

- An interval in t is defined for 11 output points, so that tout=0,0.1,...,1.

```
#
# Independent variable for ODE integration
  t0=0;tf=1;nout=11;
  tout=seq(from=t0,to=tf,by=(tf-t0)/(nout-1));
```

- Initial conditions for the two models include a cos half wave.

```
#
# Initial conditions
  u0=rep(0,3*nr*nc);
  for(ir in 1:nr){
    u0[ir]      =0.5*(1+cos(pi*(r[ir]-rl)/(ru-rl)));
    u0[ir+nr]   =0;
    u0[ir+2*nr]=0;
  }
  if(nc==2){
  for(ir in 1:nr){
    u0[ir+3*nr]=0.5*(1+cos(pi*(r[ir]-rl)/(ru-rl)));
    u0[ir+4*nr]=0;
    u0[ir+5*nr]=0;
  }
  }
  ncall=0;
```

nc is used to select the ICs for the two models. The total number of ICs is 3*nr*nc for 3 PDEs in each model, nr radial points in each PDE (the interval in the radial index is nr), and nc component models.

Also, the counter for the calls to pde1b is initialized.

- The system of $3(21) = 63$ ODEs for each PDE model is integrated by the library integrator lsodes (available in deSolve, [2]). As expected, the inputs to lsodes are the ODE/MOL function, pde1b, the IC vector u0, and the vector of output values of t, tout. The length of u0 (63 for nc=1, 126 for nc=2) informs lsodes how many ODEs are to be integrated. func,y,times are reserved names.

```
#
# ODE integration
  out=lsodes(y=u0,times=tout,func=pde1b,
      sparsetype ="sparseint",rtol=1e-6,
      atol=1e-6,maxord=5);
  nrow(out)
  ncol(out)
```

- Arrays for the numerical PDE solutions are defined as m1,c1,a1 for eqs. (6.3-1–6.3-6) and (6.4-1–6.4-3), m2,c2,a2 for eqs. (6.5-1–6.5-6) and (6.6-1–6.6-3)

```
#
# Arrays for plotting numerical solution
  m1=matrix(0,nrow=nr,ncol=nout);
  c1=matrix(0,nrow=nr,ncol=nout);
  a1=matrix(0,nrow=nr,ncol=nout);
  for(it in 1:nout){
    for(ir in 1:nr){
       m1[ir,it]=out[it,ir+1];
       c1[ir,it]=out[it,ir+1+nr];
       a1[ir,it]=out[it,ir+1+2*nr];
    }
  }
  if(nc==2){
    m2=matrix(0,nrow=nr,ncol=nout);
    c2=matrix(0,nrow=nr,ncol=nout);
    a2=matrix(0,nrow=nr,ncol=nout);
  for(it in 1:nout){
    for(ir in 1:nr){
       m2[ir,it]=out[it,ir+1+3*nr];
       c2[ir,it]=out[it,ir+1+4*nr];
       a2[ir,it]=out[it,ir+1+5*nr];
    }
  }
  }
```

nr is the interval of the index in r for the successive PDEs. The offset $+1$
is required since each ODE solution vector in out from lsodes starts with
the value of t for the solution vector.

- The numerical solutions for eqs. (6.3-1–6.3-6) and (6.4-1–6.4-3) for nc=1
and eqs. (6.5-1–6.5-6) and (6.6-1–6.6-3) for nc=2 are displayed at $t = t_0 = 0, t = t_f = 1$.

```
#
# Display numerical solution
  for(it in 1:nout){
    if((it==1)|(it==nout)){
      cat(sprintf("\n   t   r   m1(r,t)   c1(r,t)   a1(r,t)\n"));
      for(ir in 1:nr){
        cat(sprintf("%6.2f%6.2f%12.3e%12.3e%12.3e\n",
            tout[it],r[ir],m1[ir,it],c1[ir,it],a1[ir,it]));
      }
    }
  }
  if(nc==2){
  for(it in 1:nout){
    if((it==1)|(it==nout)){
      cat(sprintf("\n   t   r   m2(r,t)   c2(r,t)   a2(r,t)\n"));
```

```
        for(ir in 1:nr){
          cat(sprintf("%6.2f%6.2f%12.3e%12.3e%12.3e\n",
              tout[it],r[ir],m2[ir,it],c2[ir,it],a2[ir,it]));
        }
      }
    }
  }
```

- The number of calls to pde1b is displayed at the end of the solution.

```
#
# Calls to ODE routine
  cat(sprintf("\n\n ncall = %5d\n\n",ncall));
```

- The six PDE solutions are plotted (m1,c1,a1 for nc=1, m2,c2,a2 for nc=2).

```
#
# Plot PDE solutions
#
# m1
  par(mfrow=c(1,1));
  matplot(x=r,y=m1,type="l",xlab="r",ylab="m1(r,t)",
          xlim=c(rl,ru),lty=1,main="m1(r,t)",lwd=2,
          col="black");
  persp(r,tout,m1,theta=60,phi=45,
        xlim=c(rl,ru),ylim=c(t0,tf),zlim=c(0,1.1),
        xlab="r",ylab="t",zlab="m1(r,t)");
#
# c1
  par(mfrow=c(1,1));
  matplot(x=r,y=c1,type="l",xlab="r",ylab="c1(r,t)",
          xlim=c(rl,ru),lty=1,main="c1(r,t)",lwd=2,
          col="black");
#
# a1
  par(mfrow=c(1,1));
  matplot(x=r,y=a1,type="l",xlab="r",ylab="a1(r,t)",
          xlim=c(rl,ru),lty=1,main="a1(r,t)",lwd=2,
          col="black");
  if(nc==2){
#
# m2
  par(mfrow=c(1,1));
  matplot(x=r,y=m2,type="l",xlab="r",ylab="m2(r,t)",
          xlim=c(rl,ru),lty=1,main="m2(r,t)",lwd=2,
          col="black");
```

```
    persp(r,tout,m2,theta=60,phi=45,
        xlim=c(rl,ru),ylim=c(t0,tf),zlim=c(0,1.1),
        xlab="r",ylab="t",zlab="m2(r,t)");
#
# c2
    par(mfrow=c(1,1));
    matplot(x=r,y=c2,type="l",xlab="r",ylab="c2(r,t)",
        xlim=c(rl,ru),lty=1,main="c2(r,t)",lwd=2,
        col="black");
#
# a2
    par(mfrow=c(1,1));
    matplot(x=r,y=a2,type="l",xlab="r",ylab="a2(r,t)",
        xlim=c(rl,ru),lty=1,main="a2(r,t)",lwd=2,
        col="black");
    }
```

The ODE/MOL routine called by `lsodes` is next.

6.3.2 ODE/MOL routine

pde1b called in the main program of Listing 6.3 follows:

```
  pde1b=function(t,u,parms){
#
# Function pde1b computes the
# t derivative vectors of
# m1(r,t),c1(r,t),a1(r,t),
# m2(r,t),c2(r,t),a2(r,t)
#
# One vector to two three vectors
  m1=rep(0,nr);c1=rep(0,nr);a1=rep(0,nr);
  for(ir in 1:nr){
    m1[ir]=u[ir];
    c1[ir]=u[ir+nr];
    a1[ir]=u[ir+2*nr];
  }
  if(nc==2){
  m2=rep(0,nr);c2=rep(0,nr);a2=rep(0,nr);
  for(ir in 1:nr){
    m2[ir]=u[ir+3*nr];
    c2[ir]=u[ir+4*nr];
    a2[ir]=u[ir+5*nr];
  }
  }
```

```
#
# m1r,c1r,a1r
# m2r,c2r,a2r
  m1r=dss004(rl,ru,nr,m1);
  c1r=dss004(rl,ru,nr,c1);
  a1r=dss004(rl,ru,nr,a1);
  if(nc==2){
  m2r=dss004(rl,ru,nr,m2);
  c2r=dss004(rl,ru,nr,c2);
  a2r=dss004(rl,ru,nr,a2);
  }
#
# BCs
  m1r[1]=0;m1r[nr]=0;
  c1r[1]=0;c1r[nr]=0;
  a1r[1]=0;a1r[nr]=0;
  if(nc==2){
  m2r[1]=0;m2r[nr]=0;
  c2r[1]=0;c2r[nr]=0;
  a2r[1]=0;a2r[nr]=0
  }
#
# m1rr,c1rr,a1rr
# m2rr,c2rr,a2rr
  nl=2;nu=2;
  m1rr=dss044(rl,ru,nr,m1,m1r,nl,nu);
  c1rr=dss044(rl,ru,nr,c1,c1r,nl,nu);
  a1rr=dss044(rl,ru,nr,a1,a1r,nl,nu);
  if(nc==2){
  m2rr=dss044(rl,ru,nr,m2,m2r,nl,nu);
  c2rr=dss044(rl,ru,nr,c2,c2r,nl,nu);
  a2rr=dss044(rl,ru,nr,a2,a2r,nl,nu);
  }
#
# Functions of m1,c1,a1
  func11=rep(0,nr);
  func12=rep(0,nr);
  for(ir in 1:nr){
    func11[ir]=chi1*m1[ir]/(1+alpha1*c1[ir])^2;
    func12[ir]=m1[ir]/(1+beta1*a1[ir]^p1);
  }
  dfunc11=dss004(rl,ru,nr,func11);
#
# Functions of m2,c2,a2
  if(nc==2){
```

```
  func21=rep(0,nr);
  func22=rep(0,nr);
  for(ir in 1:nr){
    func21[ir]=chi2*m2[ir]/(1+alpha2*c2[ir])^2;
    func22[ir]=m2[ir]/(1+beta2*a2[ir]^p2);
  }
  dfunc21=dss004(rl,ru,nr,func21);
  }
#
# PDEs
  m1t=rep(0,nr);c1t=rep(0,nr);a1t=rep(0,nr);
  for(ir in 1:nr){
    if(ir==1){
      m1t[ir]=3*D1*m1rr[ir]-3*func11[ir]*c1rr[ir]+
              rm1*m1[ir]*c1[ir]*(1-m1[ir]);
      c1t[ir]=3*c1rr[ir]+func12[ir]-c1[ir];
      a1t[ir]=(1/tau1)*(3*a1rr[ir]+func12[ir]-a1[ir]);
    }
    if(ir==nr){
      m1t[ir]=D1*m1rr[ir]-func11[ir]*c1rr[ir]+
              rm1*m1[ir]*c1[ir]*(1-m1[ir]);
      c1t[ir]=c1rr[ir]+func12[ir]-c1[ir];
      a1t[ir]=(1/tau1)*(a1rr[ir]+func12[ir]-a1[ir]);
    }
    if((ir>1)&(ir<nr)){
      m1t[ir]=D1*(m1rr[ir]+(2/r[ir])*m1r[ir])-
              dfunc11[ir]*c1r[ir]-func11[ir]*c1rr[ir]-
              func11[ir]*(2/r[ir])*c1r[ir]+
              rm1*m1[ir]*c1[ir]*(1-m1[ir]);
      c1t[ir]=(c1rr[ir]+(2/r[ir])*c1r[ir])+
              func12[ir]-c1[ir];
      a1t[ir]=(1/tau1)*(a1rr[ir]+(2/r[ir])*a1r[ir]+
              func12[ir]-a1[ir]);
    }
  }
  if(nc==2){
  m2t=rep(0,nr);c2t=rep(0,nr);a2t=rep(0,nr);
  for(ir in 1:nr){
    if(ir==1){
      m2t[ir]=3*D2*m2rr[ir]-3*func21[ir]*c2rr[ir]+
              rm2*m2[ir]*c2[ir]*(1-m2[ir]);
      c2t[ir]=3*c2rr[ir]+func22[ir]-c2[ir];
      a2t[ir]=(1/tau2)*(3*a2rr[ir]+func22[ir]-a2[ir]);
    }
    if(ir==nr){
```

```
      m2t[ir]=D2*m2rr[ir]-func21[ir]*c2rr[ir]+
              rm2*m2[ir]*c2[ir]*(1-m2[ir]);
      c2t[ir]=c2rr[ir]+func22[ir]-c2[ir];
      a2t[ir]=(1/tau2)*(a2rr[ir]+func22[ir]-a2[ir]);
      }
    if((ir>1)&(ir<nr)){
      m2t[ir]=D2*(m2rr[ir]+(2/r[ir])*m2r[ir])-
              dfunc21[ir]*c2r[ir]-func21[ir]*c2rr[ir]-
              func21[ir]*(2/r[ir])*c2r[ir]+
              rm2*m2[ir]*c2[ir]*(1-m2[ir]);
      c2t[ir]=(c2rr[ir]+(2/r[ir])*c2r[ir])+
              func22[ir]-c2[ir];
      a2t[ir]=(1/tau2)*(a2rr[ir]+(2/r[ir])*a2r[ir]+
              func22[ir]-a2[ir]);
      }
    }
  }
#
# Two three vectors to one vector
  ut=rep(0,3*nr*nc);
  for(ir in 1:nr){
    ut[ir]      =m1t[ir];
    ut[ir+nr]   =c1t[ir];
    ut[ir+2*nr]=a1t[ir];
  }
  if(nc==2){
  for(ir in 1:nr){
    ut[ir+3*nr]=m2t[ir];
    ut[ir+4*nr]=c2t[ir];
    ut[ir+5*nr]=a2t[ir];
  }
  }
#
# Increment calls to pde1b
  ncall <<- ncall+1;
#
# Return derivative vector
  return(list(c(ut)));
  }
```

Listing 6.4   ODE/MOL routine for eqs. (6.3-1–6.6-3)

We can note the following details about Listing 6.4 (with some repetition of the discussion of Listing 2.2 so that this explanation is self-contained).

- The function is defined.

```
pde1b=function(t,u,parms){
#
# Function pde1b computes the
# t derivative vectors of
# m1(r,t),c1(r,t),a1(r,t),
# m2(r,t),c2(r,t),a2(r,t)
```

t is the current value of t in eqs. (6.3-1–6.6-3). u is the 63-vector of ODE/PDE dependent variables for nc=1, 126 for nc=2. parm is an argument to pass parameters to pde1b (unused, but required in the argument list). The arguments must be listed in the order stated to properly interface with lsodes called in the main program of Listing 6.3. The derivative vector of the LHS of eqs. (6.3-1–6.3-6), (6.5-1–6.5-6) is calculated and returned to lsodes as explained subsequently. The programming added for nc=2 is identified with (if(nc==2)).

- The solutions are placed in vectors m1 to a2 to facilitate the programming of eqs. (6.3-1–6.6-3).

```
#
# One vector to two three vectors
  m1=rep(0,nr);c1=rep(0,nr);a1=rep(0,nr);
  for(ir in 1:nr){
      m1[ir]=u[ir];
      c1[ir]=u[ir+nr];
      a1[ir]=u[ir+2*nr];
  }
  if(nc==2){
  m2=rep(0,nr);c2=rep(0,nr);a2=rep(0,nr);
  for(ir in 1:nr){
      m2[ir]=u[ir+3*nr];
      c2[ir]=u[ir+4*nr];
      a2[ir]=u[ir+5*nr];
  }
  }
```

nr is the interval of the index in r for the successive PDEs.

- The derivatives $\dfrac{\partial m_1}{\partial r}$ to $\dfrac{\partial a_2}{\partial r}$ are computed by dss004,

```
#
# m1r,c1r,a1r
# m2r,c2r,a2r
  m1r=dss004(rl,ru,nr,m1);
  c1r=dss004(rl,ru,nr,c1);
  a1r=dss004(rl,ru,nr,a1);
  if(nc==2){
  m2r=dss004(rl,ru,nr,m2);
  c2r=dss004(rl,ru,nr,c2);
```

```
a2r=dss004(rl,ru,nr,a2);
}
```

- BCs (6.4), (6.6-1–6.6-3) (homogeneous Neumann, no-flux BCs) are implemented.

```
#
# BCs
  m1r[1]=0;m1r[nr]=0;
  c1r[1]=0;c1r[nr]=0;
  a1r[1]=0;a1r[nr]=0;
  if(nc==2){
  m2r[1]=0;m2r[nr]=0;
  c2r[1]=0;c2r[nr]=0;
  a2r[1]=0;a2r[nr]=0
  }
```

Grid indices 1,nr correspond to $r = r_l = 0, r = r_u = 1$.
- The derivatives $\dfrac{\partial^2 m_1}{\partial r^2}$ to $\dfrac{\partial^2 a_2}{\partial r^2}$ are computed by dss044,

```
#
# m1rr,c1rr,a1rr
# m2rr,c2rr,a2rr
  nl=2;nu=2;
  m1rr=dss044(rl,ru,nr,m1,m1r,nl,nu);
  c1rr=dss044(rl,ru,nr,c1,c1r,nl,nu);
  a1rr=dss044(rl,ru,nr,a1,a1r,nl,nu);
  if(nc==2){
  m2rr=dss044(rl,ru,nr,m2,m2r,nl,nu);
  c2rr=dss044(rl,ru,nr,c2,c2r,nl,nu);
  a2rr=dss044(rl,ru,nr,a2,a2r,nl,nu);
  }
```

nl=2;nu=2; designate Neumann BCs at $r = r_l = 0, r = r_u = 1$.
- Functions of m_1 to a_2 in equations (6.3-1–6.3-6), (6.5-1–6.5-6) are implemented.

```
#
# Functions of m1,c1,a1
  func11=rep(0,nr);
  func12=rep(0,nr);
  for(ir in 1:nr){
    func11[ir]=chi1*m1[ir]/(1+alpha1*c1[ir])^2;
    func12[ir]=m1[ir]/(1+beta1*a1[ir]^p1);
  }
  dfunc11=dss004(rl,ru,nr,func11);
#
```

```
# Functions of m2,c2,a2
  if(nc==2){
  func21=rep(0,nr);
  func22=rep(0,nr);
  for(ir in 1:nr){
    func21[ir]=chi2*m2[ir]/(1+alpha2*c2[ir])^2;
    func22[ir]=m2[ir]/(1+beta2*a2[ir]^p2);
  }
  dfunc21=dss004(rl,ru,nr,func21);
  }
```

- The MOL ODEs approximating the PDEs, eqs. (6.3-1–6.6-3), are computed over the *r* domain.

```
#
# PDEs
  m1t=rep(0,nr);c1t=rep(0,nr);a1t=rep(0,nr);
  for(ir in 1:nr){
    if(ir==1){
      m1t[ir]=3*D1*m1rr[ir]-3*func11[ir]*c1rr[ir]+
              rm1*m1[ir]*c1[ir]*(1-m1[ir]);
      c1t[ir]=3*c1rr[ir]+func12[ir]-c1[ir];
      a1t[ir]=(1/tau1)*(3*a1rr[ir]+func12[ir]-a1[ir]);
    }
   if(ir==nr){
      m1t[ir]=D1*m1rr[ir]-func11[ir]*c1rr[ir]+
              rm1*m1[ir]*c1[ir]*(1-m1[ir]);
      c1t[ir]=c1rr[ir]+func12[ir]-c1[ir];
      a1t[ir]=(1/tau1)*(a1rr[ir]+func12[ir]-a1[ir]);
    }
    if((ir>1)&(ir<nr)){
      m1t[ir]=D1*(m1rr[ir]+(2/r[ir])*m1r[ir])-
              dfunc11[ir]*c1r[ir]-func11[ir]*c1rr[ir]-
              func11[ir]*(2/r[ir])*c1r[ir]+
              rm1*m1[ir]*c1[ir]*(1-m1[ir]);
      c1t[ir]=(c1rr[ir]+(2/r[ir])*c1r[ir])+
              func12[ir]-c1[ir];
      a1t[ir]=(1/tau1)*(a1rr[ir]+(2/r[ir])*a1r[ir]+
              func12[ir]-a1[ir]);
    }
  }
  if(nc==2){
  m2t=rep(0,nr);c2t=rep(0,nr);a2t=rep(0,nr);
  for(ir in 1:nr){
    if(ir==1){
      m2t[ir]=3*D2*m2rr[ir]-3*func21[ir]*c2rr[ir]+
              rm2*m2[ir]*c2[ir]*(1-m2[ir]);
```

```
      c2t[ir]=3*c2rr[ir]+func22[ir]-c2[ir];
      a2t[ir]=(1/tau2)*(3*a2rr[ir]+func22[ir]-a2[ir]);
    }
  if(ir==nr){
    m2t[ir]=D2*m2rr[ir]-func21[ir]*c2rr[ir]+
            rm2*m2[ir]*c2[ir]*(1-m2[ir]);
    c2t[ir]=c2rr[ir]+func22[ir]-c2[ir];
    a2t[ir]=(1/tau2)*(a2rr[ir]+func22[ir]-a2[ir]);
    }
  if((ir>1)&(ir<nr)){
    m2t[ir]=D2*(m2rr[ir]+(2/r[ir])*m2r[ir])-
            dfunc21[ir]*c2r[ir]-func21[ir]*c2rr[ir]-
            func21[ir]*(2/r[ir])*c2r[ir]+
            rm2*m2[ir]*c2[ir]*(1-m2[ir]);
    c2t[ir]=(c2rr[ir]+(2/r[ir])*c2r[ir])+
            func22[ir]-c2[ir];
    a2t[ir]=(1/tau2)*(a2rr[ir]+(2/r[ir])*a2r[ir]+
            func22[ir]-a2[ir]);
    }
  }
}
```

The close correspondence of the progrramming and the mathematics (PDEs) is an important feature of the MOL.

- The derivatives $\dfrac{\partial m_1}{\partial t}$ to $\dfrac{\partial a_2}{\partial t}$ approximated as MOL ODEs are placed in a single vector ut to return to lsodes for the next step along the solution.

```
#
# Two three vectors to one vector
  ut=rep(0,3*nr*nc);
  for(ir in 1:nr){
    ut[ir]      =m1t[ir];
    ut[ir+nr]   =c1t[ir];
    ut[ir+2*nr] =a1t[ir];
  }
  if(nc==2){
  for(ir in 1:nr){
    ut[ir+3*nr]=m2t[ir];
    ut[ir+4*nr]=c2t[ir];
    ut[ir+5*nr]=a2t[ir];
  }
  }
```

- The counter for the calls to pde1b is incremented and returned to the main program of Listing 6.3 by <<-.

```
#
# Increment calls to pde1b
  ncall <<- ncall+1;
```

The vector ut is returned as a list as required by lsodes.

```
#
# Return derivative vector
  return(list(c(ut)));
  }
```

The final } concludes pde1b.

The numerical and graphical output is considered next.

6.3.3 Numerical, graphical output

Abbreviated numerical output is shown in Table 6.2 for nc=2 (set in Listing 6.3).

Table 6.2

Numerical Output for eqs. (6.3-1–6.6-3)

[1] 11

[1] 127

```
   t     r     m1(r,t)      c1(r,t)      a1(r,t)
 0.00  0.00   1.000e+00    0.000e+00    0.000e+00
 0.00  0.05   9.938e-01    0.000e+00    0.000e+00
 0.00  0.10   9.755e-01    0.000e+00    0.000e+00
          .              .
          .              .
          .              .
 Output for t=0.15 to 0.85 is removed
          .              .
          .              .
          .              .
 0.00  0.90   2.447e-02    0.000e+00    0.000e+00
 0.00  0.95   6.156e-03    0.000e+00    0.000e+00
 0.00  1.00   0.000e+00    0.000e+00    0.000e+00
```

(Continued)

Table 6.2

Numerical Output for eqs. (6.3-1–6.6-3) (*Continued*)

t	r	$m1(r,t)$	$c1(r,t)$	$a1(r,t)$
1.00	0.00	2.143e-01	1.182e-01	1.182e-01
1.00	0.05	2.142e-01	1.182e-01	1.182e-01
1.00	0.10	2.141e-01	1.181e-01	1.181e-01
		.	.	
		.	.	
		.	.	

Output for t=0.15 to 0.85 is removed

		.	.	
		.	.	
		.	.	
1.00	0.90	2.058e-01	1.177e-01	1.177e-01
1.00	0.95	2.057e-01	1.177e-01	1.177e-01
1.00	1.00	2.056e-01	1.177e-01	1.177e-01

t	r	$m2(r,t)$	$c2(r,t)$	$a2(r,t)$
0.00	0.00	1.000e+00	0.000e+00	0.000e+00
0.00	0.05	9.938e-01	0.000e+00	0.000e+00
0.00	0.10	9.755e-01	0.000e+00	0.000e+00
		.	.	
		.	.	
		.	.	

Output for t=0.15 to 0.85 is removed

		.	.	
		.	.	
		.	.	
0.00	0.90	2.447e-02	0.000e+00	0.000e+00
0.00	0.95	6.156e-03	0.000e+00	0.000e+00
0.00	1.00	0.000e+00	0.000e+00	0.000e+00

t	r	$m2(r,t)$	$c2(r,t)$	$a2(r,t)$
1.00	0.00	2.143e-01	1.182e-01	1.182e-01
1.00	0.05	2.142e-01	1.182e-01	1.182e-01
1.00	0.10	2.141e-01	1.181e-01	1.181e-01
		.	.	
		.	.	
		.	.	

(*Continued*)

Table 6.2

Numerical Output for eqs. (6.3-1–6.6-3) (*Continued*)

```
Output for t=0.15 to 0.85 is removed
       .                    .
       .                    .
       .                    .
       .                    .
1.00   0.90   2.058e-01   1.177e-01   1.177e-01
1.00   0.95   2.057e-01   1.177e-01   1.177e-01
1.00   1.00   2.056e-01   1.177e-01   1.177e-01

ncall =    218
```

We can note the following details of the numerical solution in Table 6.2.

- 11 output points in t are reduced to two points in t, $t = t_0 = 0, t = t_f = 1$, from the programming in Listing 6.3.

  [1] 11

- 1+3*nr*nc = 1+3*21*2 = 127 elements in each ODE solution vector in out

  [1] 127

- The half cos ICs are confirmed for the two models from nc=2.

```
   t      r      m1(r,t)
  0.00   0.00   1.000e+00
  0.00   1.00   0.000e+00


   t      r      m2(r,t)
  0.00   0.00   1.000e+00
  0.00   1.00   0.000e+00
```

- The solutions for the two models are the same (for the same parameters in Listing 6.3). This is an important check since different solutions would indicate a programming error.
- lsodes efficiently computed a solution with ncall = 218.

The graphical output for the two models is the same as in Figs. 2.2-1a, 2.2-1b to 2.2-3 and therefore is not repeated.

To conclude this discussion of simultaneous chemotaxis inflammation models, the PDE models of eqs. (6.3-1–6.6-3) are extended to include coupling between the two models. Specifically, a source term for pro-inflammatory cytokines is postulated which is a nonlinear product function of $m_1(r,t)$ and $m_2(r,t)$. The coupled eqs. (6.3-2), (6.3-5) and (6.5-2) (6.5-5) are

$$\frac{\partial c_1}{\partial t} = \frac{\partial^2 c_1}{\partial r^2} + \frac{2}{r}\frac{\partial c_1}{\partial r} + \frac{m_1}{1+\beta_1 a_1^{p_1}} - c_1 + r_{c1}m_1 m_2 \qquad (6.7\text{-}1)$$

$$\frac{\partial c_1}{\partial t} = 3\frac{\partial^2 c_1}{\partial r^2} + \frac{m_1}{1+\beta_1 a_1^{p_1}} - c_1 + r_{c1}m_1 m_2 \qquad (6.7\text{-}2)$$

$$\frac{\partial c_2}{\partial t} = \frac{\partial^2 c_2}{\partial r^2} + \frac{2}{r}\frac{\partial c_2}{\partial r} + \frac{m_2}{1+\beta_2 a_2^{p_2}} - c_2 + r_{c2}m_1 m_2 \qquad (6.7\text{-}3)$$

$$\frac{\partial c_2}{\partial t} = 3\frac{\partial^2 c_2}{\partial r^2} + \frac{m_2}{1+\beta_2 a_2^{p_2}} - c_2 + r_{c2}m_1 m_2 \qquad (6.7\text{-}4)$$

The coupling coefficients r_{c1}, r_{c2} are defined numerically in Listing 6.3 (for nc=2).

```
#
# Number of components
  nc=2;
#
# Parameters
  D1=0.25;
  rm1=1;
  chi1=1;
  alpha1=1;
  beta1=1;
  p1=1;
  tau1=1;
  m10=1;
  rc1=1;
#
# Two components
  if(nc==2){
  D2=0.25;
  rm2=1;
  chi2=1;
  alpha2=1;
  beta2=1;
  p2=1;
  tau2=1;
  m20=1;
  rc2=10;
  }
```

The ODE/MOL routine of Listing 6.4 includes the coupling of eqs. (6.7-1–6.7-4).

```
#
# PDEs
  for(ir in 1:nr){
    if(ir==1){
      c1t[ir]=3*c1rr[ir]+func12[ir]-c1[ir]+
              rc1*m1[ir]*m2[ir];
    }
    if(ir==nr){
      c1t[ir]=c1rr[ir]+func12[ir]-c1[ir]+
              rc1*m1[ir]*m2[ir];
    }
    if((ir>1)&(ir<nr)){
      c1t[ir]=(c1rr[ir]+(2/r[ir])*c1r[ir])+
              func12[ir]-c1[ir]+
              rc1*m1[ir]*m2[ir];
    }
  }
  if(nc==2){
  for(ir in 1:nr){
    if(ir==1){
      c2t[ir]=3*c2rr[ir]+func22[ir]-c2[ir]+
              rc2*m1[ir]*m2[ir];
    if(ir==nr){
      c2t[ir]=c2rr[ir]+func22[ir]-c2[ir]+
              rc2*m1[ir]*m2[ir];
    }
    if((ir>1)&(ir<nr)){
      c2t[ir]=(c2rr[ir]+(2/r[ir])*c2r[ir])+
              func22[ir]-c2[ir]+
              rc2*m1[ir]*m2[ir];
    }
  }
  }
```

Execution of Listings 6.3 and 6.4 with these parameters (rc1, rc2) and this coding of c1t, c2t gives the following numerical output (see Table 6.3):

We can note the following details of the numerical solution in Table 6.2.

- $m_1(r,t)$ and particularly $m_2(r,t)$ are changed as expected from the values of $r_{c1} = 1, r_{c2} = 10$. In other words, $m_1(r,t)$, $m_2(r,t)$ are dependent on $c_1(r,t)$ and $c_2(r,t)$ through eqs. (6.3-1), (6.3-4) and eqs. (6.5-1), (6.5-4), respectively.

Table 6.3

Numerical Output for Eqs. (6.3-1–6.6-3) with Coupling of c1t, c2t

[1] 11

[1] 127

t	r	m1(r,t)	c1(r,t)	a1(r,t)
0.00	0.00	1.000e+00	0.000e+00	0.000e+00
0.00	0.05	9.938e-01	0.000e+00	0.000e+00
0.00	0.10	9.755e-01	0.000e+00	0.000e+00
		.	.	
		.	.	
		.	.	

Output for t=0.15 to 0.85 is removed

		.	.	
		.	.	
		.	.	
0.00	0.90	2.447e-02	0.000e+00	0.000e+00
0.00	0.95	6.156e-03	0.000e+00	0.000e+00
0.00	1.00	0.000e+00	0.000e+00	0.000e+00

t	r	m1(r,t)	c1(r,t)	a1(r,t)
1.00	0.00	2.186e-01	1.496e-01	1.190e-01
1.00	0.05	2.186e-01	1.496e-01	1.190e-01
1.00	0.10	2.184e-01	1.496e-01	1.190e-01
		.	.	
		.	.	
		.	.	

Output for t=0.15 to 0.85 is removed

		.	.	
		.	.	
		.	.	
1.00	0.90	2.087e-01	1.487e-01	1.185e-01
1.00	0.95	2.085e-01	1.487e-01	1.185e-01
1.00	1.00	2.085e-01	1.487e-01	1.185e-01

t	r	m2(r,t)	c2(r,t)	a2(r,t)
0.00	0.00	1.000e+00	0.000e+00	0.000e+00
0.00	0.05	9.938e-01	0.000e+00	0.000e+00
0.00	0.10	9.755e-01	0.000e+00	0.000e+00

(Continued)

Table 6.3

Numerical Output for Eqs. (6.3-1–6.6-3) with Coupling of `c1t,` `c2t`
(*Continued*)

```
        .                   .
        .                   .
        .                   .
Output for t=0.15 to 0.85 is removed
        .                   .
        .                   .
        .                   .
0.00  0.90    2.447e-02    0.000e+00    0.000e+00
0.00  0.95    6.156e-03    0.000e+00    0.000e+00
0.00  1.00    0.000e+00    0.000e+00    0.000e+00

   t     r     m2(r,t)      c2(r,t)      a2(r,t)
1.00  0.00    2.571e-01    4.328e-01    1.268e-01
1.00  0.05    2.570e-01    4.327e-01    1.268e-01
1.00  0.10    2.565e-01    4.326e-01    1.268e-01
        .                   .
        .                   .
        .                   .
Output for t=0.15 to 0.85 is removed
        .                   .
        .                   .
        .                   .
1.00  0.90    2.367e-01    4.277e-01    1.257e-01
1.00  0.95    2.364e-01    4.276e-01    1.257e-01
1.00  1.00    2.363e-01    4.276e-01    1.257e-01

ncall =    261
```

```
    Table 6.2

       t     r     m1(r,t)      c1(r,t)      a1(r,t)
     1.00  0.00    2.143e-01    1.182e-01    1.182e-01

     1.00  1.00    2.056e-01    1.177e-01    1.177e-01

       t     r     m2(r,t)      c2(r,t)      a2(r,t)
     1.00  0.00    2.143e-01    1.182e-01    1.182e-01

     1.00  1.00    2.056e-01    1.177e-01    1.177e-01
```

Table 6.3

t	r	m1(r,t)	c1(r,t)	a1(r,t)
1.00	0.00	2.186e-01	1.496e-01	1.190e-01
1.00	1.00	2.085e-01	1.487e-01	1.185e-01

t	r	m2(r,t)	c2(r,t)	a2(r,t)
1.00	0.00	2.571e-01	4.328e-01	1.268e-01
1.00	1.00	2.363e-01	4.276e-01	1.257e-01

- The preceding comparison (Tables 6.2 and 6.3) indicates $c_1(r,t)$ and particularly $c_2(r,t)$ are changed as expected from the values of $r_{c1} = 1, r_{c2} = 10$ through eqs. (6.7-1–6.7-4).
- The preceding comparison (Tables 6.2 and 6.3) indicates $a_1(r,t)$ and $a_2(r,t)$ are changed as expected from the values of $r_{c1} = 1, r_{c2} = 10$. In other words, $a_1(r,t)$, $a_2(r,t)$ are dependent on $m_1(r,t)$ and $m_2(r,t)$ through eqs. (6.3-3), (6.3-6) and eqs. (6.5-3), (6.5-6), respectively.
- lsodes efficiently computed the solution of Table 6.3 with ncall = 261.

In summary, the solutions in Table 6.3 illustrate the complex interconnection of eqs. (6.3-1–6.6-3) from just the terms with r_{c1}, r_{c2} in eqs. (6.7-1–6.7-4).

The graphical output from the extensions of Listings 6.3 and 6.4 is as follows:

Figures 6.2-1a, 6.2-1b can be compared with Figure 2.2-1a, 2.2-1b to observe the changes in $m_1(r,t)$ resulting from the cross coupling of eqs. (6.7-1–6.7-4).

Figures 6.2-2 and 6.2-3 can be compared with Figures 2.2-2 and 2.2-3 to observe the changes in $c_1(r,t)$, $a_1(r,t)$ resulting from the cross coupling of eqs. (6.7-1–6.7-4).

Figure 6.2-4a, 6.2-4b can be compared with Figures 2.2-1a, 2.2-1b to observe the changes in $m_2(r,t)$ resulting from the cross coupling of eqs. (6.7-1–6.7-4).

Figures 6.2-5, 6.2-6 can be compared with Figures 2.2-2, 2.2-3 to observe the changes in $c_2(r,t)$, $a_2(r,t)$ resulting from the cross coupling of eqs. (6.7-1–6.7-4).

This concludes the development and discussion of eqs. (6.3-1–6.7-4) illustrating two coupled chemotaxis inflammation models.

m1(r,t)

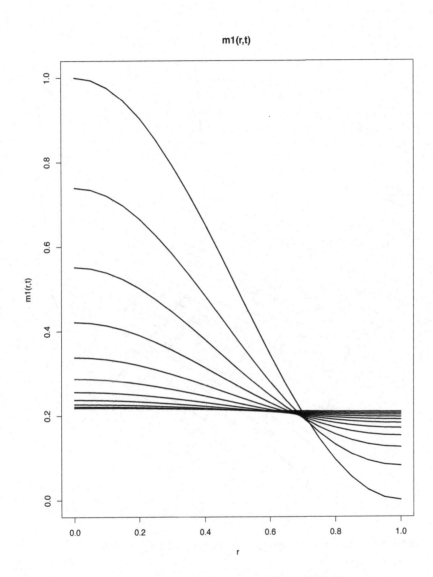

Figure 6.2-1a Numerical solution $m_1(r,t)$ from eqs. (6.3-1–6.7-4), 2D.

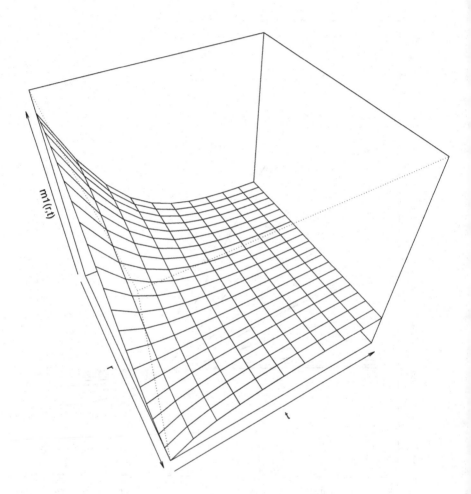

Figure 6.2-1b   Numerical solution $m_1(r,t)$ from eqs. (6.3-1–6.7-4), 3D.

c1(r,t)

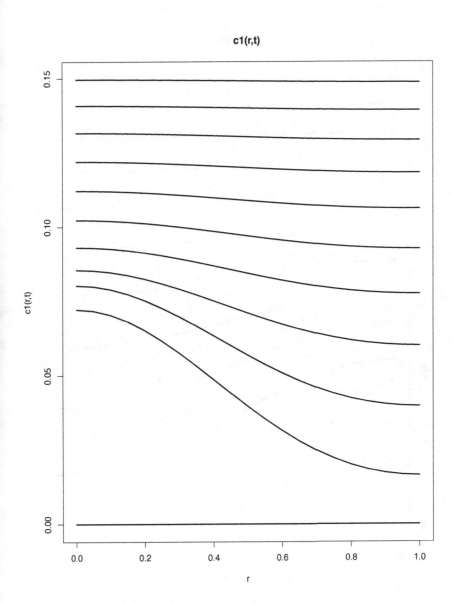

Figure 6.2-2   Numerical solution $c_1(r,t)$ from eqs. (6.3-1–6.7-4), 2D.

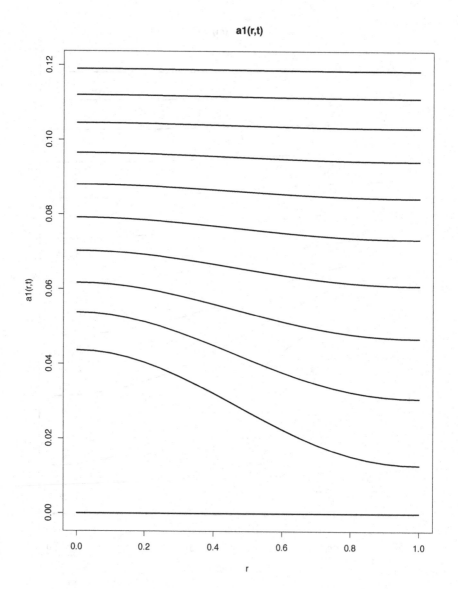

Figure 6.2-3 Numerical solution $a_1(r,t)$ from eqs. (6.3-1–6.7-4).

m2(r,t)

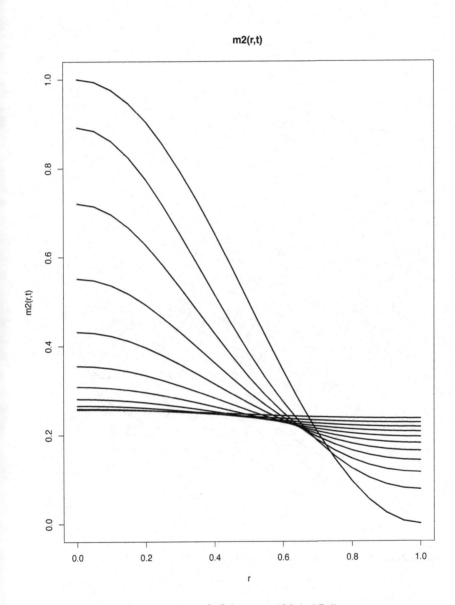

Figure 6.2-4a Numerical solution $m_2(r,t)$ from eqs. (6.3-1–6.7-4).

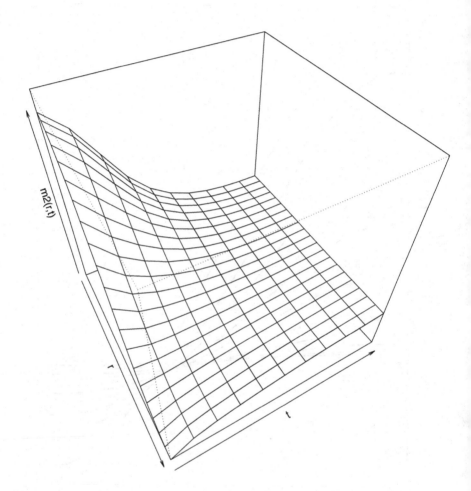

Figure 6.2-4b   Numerical solution $m_2(r,t)$ from eqs. (6.3-1–6.7-4), 3D.

c2(r,t)

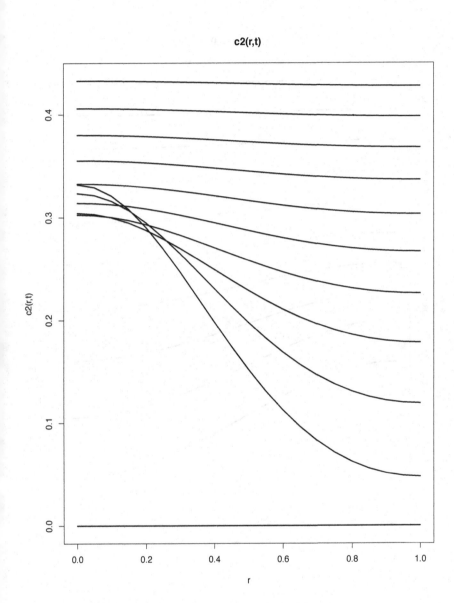

Figure 6.2-5 Numerical solution $c_2(r,t)$ from eqs. (6.3-1–6.7-4).

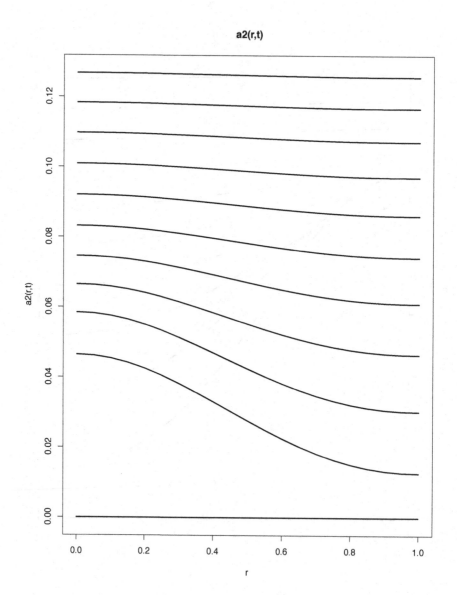

Figure 6.2-6   Numerical solution $a_2(r,t)$ from eqs. (6.3-1–6.7-4).

Summary and conclusion

Two case studies of the chemotaxis inflammation model are considered in this chapter: (1) variation/sensitivity of parameters, and (2) coupling of two models. The changes in the main program and ODE/MOL routines are straightforward and illustrate a methodology for parameter and coupled models development, as well as the extension to other inflammation models that might be suggested as a result of experimental/observational studies, e.g., Covid-19 neurological effects (long Covid).

References

1. Giunta, M., Carmela Lombardo, and M. Sammartino (2021), Pattern Formation and Transition to Chaos in a Chemotaxis Model of Acute Inflammation. *SIAM Journal on Applied Dynamical Systems*, **20**, no. 4, pp. 1844–1881.

2. Soetaert, K., J. Cash, and F. Mazzia (2012), *Solving Differential Equations in R*, Springer-Verlag, Heidelberg, Germany.

A1 Functions dss004, dss044

A1.1 dss004 **listing**

A listing of function dss004 is as follows:

```
  dss004=function(xl,xu,n,u) {
#
# An extensive set of documentation comments detailing
# the derivation of the following fourth order finite
# differences (FDs) is not given here to conserve
# space. The derivation is detailed in Schiesser,
# W. E., The Numerical Method of Lines Integration
# of Partial Differential Equations, Academic Press,
# San Diego, 1991.
#
# Preallocate arrays
  ux=rep(0,n);
#
# Grid spacing
  dx=(xu-xl)/(n-1);
#
# 1/(12*dx) for subsequent use
  r12dx=1/(12*dx);
#
# ux vector
#
# Boundaries (x=xl,x=xu)
  ux[1]=r12dx*(-25*u[1]+48*u[ 2]-36*u[ 3]+16*u[ 4]-3*u[ 5]);
  ux[n]=r12dx*( 25*u[n]-48*u[n-1]+36*u[n-2]-16*u[n-3]+3*u[n-4]);
#
# dx in from boundaries (x=xl+dx,x=xu-dx)
  ux[ 2]=r12dx*(-3*u[1]-10*u[ 2]+18*u[ 3]-6*u[ 4]+u[ 5]);
  ux[n-1]=r12dx*( 3*u[n]+10*u[n-1]-18*u[n-2]+6*u[n-3]-u[n-4]);
#
# Interior points (x=xl+2*dx,...,x=xu-2*dx)
  for(i in 3:(n-2))ux[i]=r12dx*(-u[i+2]+8*u[i+1]-8*u[i-1]+u[i-2]);
#
# All points concluded (x=xl,...,x=xu)
  return(c(ux));
}
```
Listing A1.1 dss004

DOI: 10.1201/9781003311201-A1

We can note the following details about Listing A1.1:

The input arguments are

> xl   lower boundary value of x
>
> xu   upper boundary value of x
>
> n   number of points in the grid in x, including the end points
>
> u   dependent variable to be differentiated, an n-vector

The output, ux, is an n-vector of numerical values of the first derivative of u.

The finite difference (FD) approximations are a weighted sum of the dependent variable values. For example, at point i

```
for(i in 3:(n-2))ux[i]=r12dx*(-u[i+2]+8*u[i+1]-8*u[i-1]+u[i-2]);
```

The weighting coefficients are −1, 8, 0, −8, 1 at points i-2, i-1, i, i+1, i+2, respectively. These weighting coefficients are antisymmetric (opposite sign) around the center point i because the computed first derivative is of odd order. If the derivative is of even order, the weighting coefficients would be symmetric (same sign) around the center point.

For i=1, the dependent variable at points i=1,2,3,4,5 is used in the FD approximation for ux[1] to remain within the x domain (fictitious points outside the x domain are not used).

```
ux[1]=r12dx*(-25*u[1]+48*u[2]-36*u[3]+16*u[4]-3*u[5]);
```

Similarly, for i=2, points i=1,2,3,4,5 are used in the FD approximation for ux[2] to remain within the x domain (fictitious points outside the x domain are avoided).

```
ux[2]=r12dx*(-3*u[1]-10*u[2]+18*u[3]-6*u[4]+u[5]);
```

At the right boundary $x = x_u$, points at i=n,n-1,n-2,n-3,n-4 are used for ux[n],ux[n-1] to avoid points outside the x domain.

In all cases, the FD approximations are fourth order correct in x.

A1.2   dss044 listing

A listing of function dss044 is as follows:

```
dss044=function(xl,xu,n,u,ux,nl,nu) {
#
# The derivation of the finite difference
```

```
# approximations for a second derivative are
# in Schiesser, W. E., The Numerical Method
# of Lines Integration of Partial Differential
# Equations, Academic Press, San Diego, 1991.
#
# Preallocate arrays
  uxx=rep(0,n);
#
# Grid spacing
  dx=(xu-xl)/(n-1);
#
# 1/(12*dx**2) for subsequent use
  r12dxs=1/(12*dx^2);
#
# uxx vector
#
# Boundaries (x=xl,x=xu)
  if(nl==1)
    uxx[1]=r12dxs*
          (45*u[ 1]-154*u[ 2]+214*u[ 3]-
          156*u[ 4] +61*u[ 5] -10*u[ 6]);
  if(nu==1)
    uxx[n]=r12dxs*
          (45*u[ n]-154*u[n-1]+214*u[n-2]-
          156*u[n-3] +61*u[n-4] -10*u[n-5]);
  if(nl==2)
    uxx[1]=r12dxs*
          (-415/6*u[ 1] +96*u[ 2]-36*u[ 3]+
            32/3*u[ 4]-3/2*u[ 5]-50*ux[1]*dx);
  if(nu==2)
    uxx[n]=r12dxs*
          (-415/6*u[ n] +96*u[n-1]-36*u[n-2]+
            32/3*u[n-3]-3/2*u[n-4]+50*ux[n]*dx);
#
# dx in from boundaries (x=xl+dx,x=xu-dx)
    uxx[ 2]=r12dxs*
          (10*u[ 1]-15*u[ 2]-4*u[ 3]+
          14*u[ 4]- 6*u[ 5] +u[ 6]);
    uxx[n-1]=r12dxs*
          (10*u[ n]-15*u[n-1]-4*u[n-2]+
          14*u[n-3]- 6*u[n-4] +u[n-5]);
#
# Remaining interior points (x=xl+2*dx,...,
# x=xu-2*dx)
  for(i in 3:(n-2))
```

```
    uxx[i]=r12dxs*
        (-u[i-2]+16*u[i-1]-30*u[i]+
        16*u[i+1]   -u[i+2]);
#
# All points concluded (x=xl,...,x=xu)
  return(c(uxx));
}
```

Listing A1.2 dss044

We can note the following details about Listing A1.2:

The input arguments are

xl   lower boundary value of x

xu   upper boundary value of x

n   number of points in the grid in x,
      including the end points

u   dependent variable to be differentiated,
      an n-vector

ux   first derivative of u with boundary
      condition (BC) values, an n-vector

nl   type of boundary condition at x=xl
      1: Dirichlet BC
      2: Neumann BC

nu   type of boundary condition at x=xu
      1: Dirichlet BC
      2: Neumann BC

The output, uxx, is an n-vector of numerical values of the second derivative of u.

The finite difference (FD) approximations are a weighted sum of the dependent variable values. For example, at point i

```
  for(i in 3:(n-2))
    uxx[i]=r12dxs*
        (-u[i-2]+16*u[i-1]-30*u[i]+
        16*u[i+1]   -u[i+2]);
```

The weighting coefficients are -1, 16, -30, 16, -1 at points i-2, i-1, i, i+1, i+2, respectively. These weighting coefficients are symmetric around the center point i because the computed second derivative is of even order. If the derivative is of odd order, the weighting coefficients would be antisymmetric (opposite sign) around the center point.

For nl=2 and/or nu=2, the boundary values of the first derivative are included in the FD approximation for the second derivative, uxx. For example, at x=xl (with nl=2),

```
if(nl==2)
  uxx[1]=r12dxs*
            (-415/6*u[  1] +96*u[  2]-36*u[  3]+
             32/3*u[  4]-3/2*u[  5]-50*ux[1]*dx);
```

In computing the second derivative at the left boundary, uxx[1], the first derivative at the left boundary is included, that is, ux[1]. In this way, a Neumann BC is accommodated (ux[1] is included in the input argument ux).

For nl=1, only values of the dependent variable (and not the first derivative) are included in the weighted sum.

```
if(nl==1)
  uxx[1]=r12dxs*
            (45*u[  1]-154*u[  2]+214*u[  3]-
             156*u[  4] +61*u[  5] -10*u[  6]);
```

The dependent variable at points i=1,2,3,4,5,6 is used in the FD approximation for uxx[1] to remain within the x domain (fictitious points outside the x domain are not used).

Six points are used rather than five (as in the centered approximation for uxx[i]) since the FD applies at the left boundary and is not centered (around i). Six points provide a fourth order FD approximation which is the same order as the FDs at the interior points in x.

Similar considerations apply at the upper boundary value of x with nu=1,2.

Robin boundary conditions can also be accommodated with nl=2, nu=2. In all three cases, Dirichlet, Neumann and Robin, the boundary conditions can be linear and/or nonlinear.

A2 Accuracy of Numerical PDE Solutions

For most realistic PDE applications, typically modeled by systems of nonlinear PDEs, analytical (mathematical, exact) solutions are generally unavailable. Alternatively, numerical solutions can, in principle, be computed for PDE models of any order (number of PDEs) and nonlinearity. However, the question of the accuracy of the numerical solutions should be addressed. In this appendix, two methods are discussed for assessing the accuracy of numerical PDE solutions that do not require an analytical solution.

A2.1 h refinement

Truncation error is the principal error in computing numerical PDE solutions by finite differences (FDs). Briefly, this is the error resulting from truncation of the Taylor series that is the basis of FD methods. Truncation error is of the form

$$error = O(h^p) \tag{A2.1}$$

where O denotes *of order*, h is the FD spacing, and p is the order of the FD approximation. Equation (A2.1) indicates that $error \to 0$ as $h \to 0$. To test this idea of convergence, the number of radial grid points nr in the main program and ODE/MOL routine of Listings 2.1 and 2.2 can be increased from nr=21 (so that h is reduced). The abbreviated solution for nr=21,41,81 is tabulated next (ncase=2).

```
nr=21 (Table 2.2)

   t      r      m(r,t)      c(r,t)      a(r,t)
 1.00   0.00    2.143e-01   1.182e-01   1.182e-01

 1.00   1.00    2.056e-01   1.177e-01   1.177e-01

 ncall =    155


nr=41

   t      r      m(r,t)      c(r,t)      a(r,t)
 1.00   0.00    2.143e-01   1.182e-01   1.182e-01

 1.00   1.00    2.056e-01   1.177e-01   1.177e-01
```

DOI: 10.1201/9781003311201-A2

```
ncall =    229


nr=81


   t     r      m(r,t)       c(r,t)        a(r,t)
1.00  0.00    2.143e-01    1.182e-01     1.182e-01


1.00  1.00    2.056e-01    1.177e-01     1.177e-01


ncall =    348
```

These results imply four figure convergence of the numerical solution for nr=21. In other words, the previously computed solutions of the chemotaxis inflammation model have sufficient accuracy in the related discussion. The computational effort increases with the increasing number of radial points nr as expected.

There is the question of the accuracy of the PDE solutions with respect to t. This could be investigated by varying the error tolerances in the calls to lsodes. This is left as an exercise.

In summary, the apparent convergence of the FD MOL solutions of the PDE models discussed previously has been established. This is not a proof of accuracy, but only an indication of the apparent accuracy based on the FD approximations in dss004, dss044 for the numerical integration in r, and the ODE/MOL solutions from lsodes for the numerical integration in t. The usual practice is to accept the FD solution after completing an h refinement analysis, and possibly a convegence analysis of the t integration, e.g. , by varying the lsodes error tolerances.

In general, this approach to h refinement requires that the numerical solution is: (1) stable, (2) convergent, and (3) consistent (converges to the solution of the correct PDE).

A2.2   p refinement

The order of the FD approximation in eq. (A2.1) is p, and variations in this order to determine the convergence of the numerical solution is therefore termed as *p refinement*. dss004, dss044 used in Listings 2.1 and 2.2 is fourth order, $p = 4$. This value can be changed by calling FD routines of different orders. For example, the FDs in dss006, dss046 is sixth order, $p = 6$.

dss006, dss046 can be called in place of dss004, dss044 by just changing the routine names since the arguments are the same. Therefore, the changes in the main program of Listing 2.1 are

```
#
# Access functions for numerical solution
  setwd("f:/inflammation/chap2");
  source("pde1a.R");
  source("dss006.R");
  source("dss046.R");
```

The changes to the ODE/MOL routine in Listing 2.2 are dss004, dss044 to dss006, dss046.

```
#
# mr,cr,ar
   mr=dss004(rl,ru,nr,m);
   cr=dss004(rl,ru,nr,c);
   ar=dss004(rl,ru,nr,a);
#
# BCs
   mr[1]=0;mr[nr]=0;
   cr[1]=0;cr[nr]=0;
   ar[1]=0;ar[nr]=0;
#
# mrr,crr,arr
   nl=2;nu=2;
   mrr=dss044(rl,ru,nr,m,mr,nl,nu);
   crr=dss044(rl,ru,nr,c,cr,nl,nu);
   arr=dss044(rl,ru,nr,a,ar,nl,nu);
#
# Functions of m,c,a
   func1=rep(0,nr);
   func2=rep(0,nr);
   for(ir in 1:nr){
     func1[ir]=chi*m[ir]/(1+alpha*c[ir])^2;
     func2[ir]=m[ir]/(1+beta*a[ir]^p);
   }
   dfunc1=dss004(rl,ru,nr,func1);
```

   changed to

```
#
# mr,cr,ar
   mr=dss006(rl,ru,nr,m);
   cr=dss006(rl,ru,nr,c);
   ar=dss006(rl,ru,nr,a);
#
# BCs
   mr[1]=0;mr[nr]=0;
   cr[1]=0;cr[nr]=0;
   ar[1]=0;ar[nr]=0;
#
# mrr,crr,arr
   nl=2;nu=2;
   mrr=dss046(rl,ru,nr,m,mr,nl,nu);
   crr=dss046(rl,ru,nr,c,cr,nl,nu);
```

```
  arr=dss046(rl,ru,nr,a,ar,nl,nu);
#
# Functions of m,c,a
  func1=rep(0,nr);
  func2=rep(0,nr);
  for(ir in 1:nr){
    func1[ir]=chi*m[ir]/(1+alpha*c[ir])^2;
    func2[ir]=m[ir]/(1+beta*a[ir]^p);
  }
  dfunc1=dss006(rl,ru,nr,func1);
```

The output for dss004, dss044 (p=4), dss006, dss046 (p=6), dss008, dss048 (p=8) is as follows:

```
dss004, dss044 (Table 2.2), p=4
```

t	r	m(r,t)	c(r,t)	a(r,t)
1.00	0.00	2.143e-01	1.182e-01	1.182e-01
1.00	1.00	2.056e-01	1.177e-01	1.177e-01

```
ncall =    155
```

```
dss006, dss046, p=6
```

t	r	m(r,t)	c(r,t)	a(r,t)
1.00	0.00	2.143e-01	1.182e-01	1.182e-01
1.00	1.00	2.056e-01	1.177e-01	1.177e-01

```
ncall =    182
```

```
dss008, dss048, p=8
```

t	r	m(r,t)	c(r,t)	a(r,t)
1.00	0.00	2.143e-01	1.182e-01	1.182e-01
1.00	1.00	2.056e-01	1.177e-01	1.177e-01

```
ncall =    184
```

This output indicates four figure convergence with p refinement.

The increase in the order of FD approximations is achieved by including additional points in the approroximation. For example, at the interior point i

```
dss044 (Appendix A1)
```

```
uxx[i]=r12dxs*
        (-u[i-2]+16*u[i-1]-30*u[i]+16*u[i+1]-u[i+2]);
```

dss046

```
uxx[i]=rdxs*
(    0.011111111111111*u[i-3]-0.150000000000000*u[i-2]+
     1.500000000000000*u[i-1]+0.011111111111111*u[i+3]-
     0.150000000000000*u[i+2]+1.500000000000000*u[i+1]-
     2.722222222222220*u[i  ]);
```

dss048

```
uxx[i]=rdxs*
    (-0.001785714285714*u[i-4]+0.025396825396825*u[i-3]-
     0.200000000000000*u[i-2]+1.600000000000000*u[i-1]-
     0.001785714285714*u[i+4]+0.025396825396825*u[i+3]-
     0.200000000000000*u[i+2]+1.600000000000000*u[i+1]-
     2.847222222222221*u[i  ]);
```

5,7,9 points are included for $p = 4,6,8$. These FDs indicate that additional computation is required to achieve higher order.

Index

Printed in the United States
by Baker & Taylor Publisher Services